ranfield
UNIVERSITY

vice

Hypervelocity Gouging Impacts

John D. Cinnamon
Air Force Institute of Technology
Wright-Patterson Air Force Base
Dayton, Ohio

Volume 228
PROGRESS IN
ASTRONAUTICS AND AERONAUTICS

Frank K. Lu, Editor-in-Chief
University of Texas at Arlington
Arlington, Texas

Published by the
American Institute of Aeronautics and Astronautics, Inc.
1801 Alexander Bell Drive, Reston, Virginia 20191-4344

American Institute of Aeronautics and Astronautics, Inc., Reston, Virginia

1 2 3 4 5

ISBN 978-1-56347-965-6

"Fear not, I am with you; be not dismayed; I am your God.
I will strengthen you, and help you, and uphold you with my
right hand...."

Isaiah 41:10

To Tamara, Daniel, Emily, Amy, and Jack, for your love and support.

Table of Contents

Preface

Hypervelocity impact is an area of extreme interest in the research community. The U.S. Air Force has a test facility at Holloman Air Force Base (AFB) that specializes in hypervelocity impact testing. This Holloman AFB High Speed Test Track (HHSTT) is currently working toward a test vehicle speed above Mach 10. As the sled's speed has increased to Mach 8.5, a material interaction that causes "gouging" in the rails or the sled's "shoes" has developed, and this can result in catastrophic failure.

Previous efforts in investigating this event have resulted in a choice of the most suitable computer code (CTH) and a model of the shoe/rail interaction. However, the dynamic stress models of the specific materials were not developed, and the model was not validated against experimentation.

In this work, a summary of past and present research efforts, as well as the theoretical foundation of this field of study, is presented. A characterization of gouging is developed from an examination of a gouged rail from the HHSTT. A thermodynamic history of gouging is determined from the experimental evidence, and an extensive study that determines the specific material models is performed.

The developed material dynamic strength models are validated utilizing several experimental tests that are successfully simulated using CTH. Additionally, a penetration theory that provides insight into the gouging problem using an analytic approach which does not require the use of computationally intensive codes is developed.

Based on the detailed examination of the materials and the validation of the material models within CTH, an evaluation of the HHSTT gouging phenomenon is performed. These simulations of the gouging problem replicate the experimentally observed characteristics and lead to recommendations to mitigate the occurrence of hypervelocity gouging.

John D. Cinnamon
March 2009

Acknowledgments

I would like to acknowledge the support and genuine mentorship of individuals critical to the success of this work. In particular, Anthony Palazotto's guidance has been essential. William Baker was instrumental in applying rigorous mathematical tools to the effort. Robert Brockman asked key questions and aided immensely. Michael Hooser kept the focus on finding answers and kept us on track while providing a wealth of information, ideas, and support. I simply could not have asked for a more supportive, demanding, and talented research team.

I would also like to acknowledge the expertise of Theodore Nicholas and Singh Brar, both who provided critical assistance in the experimental process. Another crucial contribution to this effort was made by the Air Force Research Laboratories, Materials Directorate. In addition, I am indebted to Scott Furlow at the HHSTT for his constant help in getting experimental materials machined. This research would not have been possible without the amazing support and encouragement of the Air Force Office of Scientific Research and the Air Force Institute of Technology (AFIT). In particular, the Department of Aeronautics and Astronautics at AFIT provided unparalleled aid in this effort.

I would also like to acknowledge the support and encouragement of Stan Jones. His positive impact early in my academic career was immeasurable, and his mentorship was critical to my pursuit of graduate education. Stan introduced me to this topic area, and taught me how to conduct rigorous scientific research.

The encouragement and sound foundation provided by my parents was instrumental in all my success. Of course, none of this would have been possible without the amazing support and love of my wife, Tamara, and my children: Daniel, Emily, Amy, and Jack. They are the true measure of my success, and their sacrifices were many.

Nomenclature

A = Hopkinson bar cross-sectional area
A = instantaneous penetrator tip cross-sectional area
A = Johnson–Cook coefficient
A = Zerilli–Armstrong coefficient
A_i = rod cross-sectional area
A_s = gauge cross-sectional area
A_4 = undamaged rail specimen
A_0 = initial penetrator tip cross-sectional area
A_1 = penetrator tip cross-sectional area at end of transient
A_1 = target cross-sectional area
a = slope of crater volume/kinetic energy curve
a_x = acceleration in the x direction
B = Johnson–Cook coefficient
B_1 = gouge specimen
B_2 = gouge specimen
B_3 = gouge specimen
B_4 = gouge specimen
b = intercept of crater volume/kinetic-energy curve
C = arbitrary constant
C = Johnson–Cook coefficient
$[C]$ = viscous damping matrix
C' = arbitrary constant
C_f = empirical friction coefficient
C_v = specific heat at constant volume
c = plastic wave speed
c_0 = elastic wave speed (speed of sound in material)
c_0 = material sound speed
c_1 = Zerilli–Armstrong coefficient
c_2 = Zerilli–Armstrong coefficient
c_3 = Zerilli–Armstrong coefficient
c_4 = Zerilli–Armstrong coefficient
c_5 = Zerilli–Armstrong coefficient
$\boldsymbol{D}$ = deviatoric stress tensor
D_f = final Taylor specimen diameter
D_0 = initial Taylor specimen diameter

d = height
d_k = power fundamental dimensions are raised to
E = elastic modulus
E = energy
E = specific internal energy
E = test material elastic modulus
E_H = entropy of the shocked material
E_0 = elastic modulus
E_0 = impact kinetic energy
e = engineering strain of penetrator head
e = specific total energy
e_0 = engineering strain at impact
e_1 = engineering strain at end of transient
F = force
$F(t)$ = forcing function
F_f = force of friction
F_I = force of impact
F_x = force of deformation in x direction
$f(x)$ = initial thermal profile
f_i = external body forces per unit mass
G_0 = shear modulus
h_f = final Taylor specimen undeformed section length
K = thermal conductivity
k = microstructural stress density
k_{linear} = linear stiffness matrix
$k_{\text{nonlinear}}$ = nonlinear stiffness matrix
L = length
L = length of the specimen test section
L_f = final Taylor specimen diameter
L_j = fundamental dimensions
L_0 = initial Taylor specimen length
l = length
ℓ = average grain diameter
ℓ = rod undeformed section length
M = mass
$[M]$ = mass matrix
$M(x)$ = Mach number as a function of the downstream distance x
M_∞ = Mach number at the free stream
m = Johnson–Cook coefficient
m = mass of projectile
m = number of dimensioned quantities
N = normal force
n = Johnson–Cook coefficient
n = Zerilli–Armstrong coefficient
P = average pressure on penetrator tip
P = forces at specimen ends
P = hydrostatic pressure
P = pressure

P_H = pressure of the shocked material
P_0 = average pressure on penetrator tip at impact
P_1 = average pressure on penetrator tip at end of transient
p = pressure
p_a = pressure on the axis of the penetrator tip
p_0 = pressure on penetrator tip axis at impact
p_1 = axial penetrator tip pressure at end of transient
p_∞ = pressure at the free stream
q_i = dimensioned quantities
R_t = target dynamic yield strength
S = entropy
$\mathbf{S}$ = spherical stress tensor
s_{ij} = stress deviator tensor
T = temperature
T = time
T^* = homologous temperature
T_{melt} = melt temperature
t = time
U_S = shock wave velocity, disturbed material
u = displacement
u = particle velocity
u = penetration velocity
$\{u\}$ = displacement vector
$\{\dot{u}\}$ = velocity vector
$\{\ddot{u}\}$ = acceleration vector
u_P = particle velocity behind shock front
u_s = shock wave speed in target material
u_x = horizontal velocity
u_y = vertical velocity
V_c = crater volume
V_s = striker bar velocity
v = current undeformed section velocity
v = sled velocity
v_i = velocity
v_0 = initial Taylor specimen velocity
v_0 = rod impact velocity
Y_p = penetrator dynamic yield strength
z = penetration depth
α_i = exponent in Buckingham pi formulation
γ = ratio of the specific heats
Δs_x = deformation in the x direction
Δt_x = duration of impact in the x direction
Δt_y = duration of impact in vertical direction
Δv_x = change of velocity in the x direction
Δv_y = impact velocity
$\Delta\sigma'_G$ = stress, related to dislocation density
ε = equivalent plastic strain
ε = strain

ε = uniaxial strain
$\dot{\varepsilon}$ = effective plastic strain-rate
$\dot{\varepsilon}^*$ = dimensionless strain rate
ε_i = incident strain pulse
ε_{ij} = strain tensor
$\dot{\varepsilon}_{ij}$ = strain-rate tensor
ε_r = reflected strain pulse
ε_t = transmitted strain pulse
μ = coefficient of friction
ν = Poisson's ratio
Π = Buckingham pi invariant
ρ = density
ρ = density of the shocked material
ρ = material density
ρ = penetrator density
ρ_t = target material density
σ = material equivalent flow stress
σ = stress
σ = uniaxial stress
σ_b = stress in the bar
σ_{ij} = stress tensor
$\sigma_{y,c}$ = compressive yield strength
σ_1 = principle stress
σ_2 = principle stress
σ_3 = principle stress

Abbreviations and Acronyms

AFB = Air Force base
AFOSR = Air Force Office of Scientific Research
AFRL = Air Force Research Laboratory
BCC = body-centered cubic
BCT = body-centered tetragonal
BSE = backscatter emitter
CTH = Chi to the Three Halves, Eulerian Shock-Wave Physics Code
DADS = Dynamic Analysis and Design System
DoD = Department of Defense
EDS = energy dispersive x-ray spectroscopy
EOS = equation of state
FCC = face-centered cubic
FLT = force, length, time
HEL = Hugoniot elastic limit
HHSTT = Holloman AFB High Speed Test Track
HRC = Rockwell hardness test C
LMI = low mass interceptor
MLT = mass, length, time
MMI = medium mass interceptor
OIM = orientation imaging spectroscopy
PIT = parallel impact thermodynamics
PMMA = polymethyl methacryate
RHA = rolled homogeneous armor
SEM = scanning electron microscope
SHB = split Hopkinson bar
UDRI = University of Dayton Research Institute
USAF = U.S. Air Force

Chapter 1

Hypervelocity Gouging Problem Overview

I. Introduction

GOUGING resulting from hypervelocity impact is an area of interest for the United States Air Force (USAF) and the Department of Defense (DoD). The Holloman High Speed Test Track (HHSTT) serves as the premier USAF facility conducting hypervelocity impact tests. The Air Force Research Laboratory (AFRL) and the Air Force Office of Scientific Research (AFOSR) are interested in increasing the capability of this facility up to the approximately Mach 10. Indeed, this research is being conducted under AFOSR sponsorship.

II. HHSTT Rocket Sled System

The HHSTT rocket sled system appears in Fig. 1.1. This particular sled is the 192-pound Missile Defense Agency payload tested on 29 April 2003. This fully instrumented test run served to validate the HHSTT hypersonic capabilities and also set the World Land Speed Record of Mach 8.5 (6416 mph or 9410 fps). The rocket sled in Fig. 1.1 rides along the standard railroad rails constructed from 1080 steel on "shoes" or "slippers" fabricated of VascoMax 300 steel. Figure 1.2 is a schematic of the interface between the sled's shoes and the rail. The shoe is formed to wrap around the rail head, leaving a very small gap (on the order of 0.5 cm). The nominal sled configuration is the one in Fig. 1.1. This study will be limited to this geometry.

The goal of the HHSTT's current efforts is to increase their velocity capability to approximately Mach 10 ($\sim$10,000 fps or $\sim$3 km/s). However, at velocities of approximately 1.5 km/s a phenomenon known as "gouging" occurs. This phenomenon is characterized by either damage (i.e., material removal and/or melting) to the rail and/or shoe or catastrophic failure of the rocket sled system caused by material interaction between the rail and the shoe. As mentioned earlier, the shoes are machined to allow a slight gap between their structure and the rail head. This allows the vehicle to maintain a limited free-flight condition as the vehicle accelerates down the track. The consequence of this free-flight condition is that the shoe can roll, pitch, or yaw with relation to the rail during the test, and this results in intermittent contact between the shoe and the rail. This contact typically creates zones of material removal observed on the rail. Because of the method of

Fig. 1.1 Rocket sled system at the HHSTT.

sled system braking, the shoes are dramatically altered in the slow-down process and therefore are not typically suitable for posttest analysis. Although noncatastrophic gouging is costly (in terms of rail repair), it does not necessarily adversely affect the test mission. However, some gouging events either destroy the rail and cause a sled crash, or multiple gouges deteriorate the sled's shoes to the point of catastrophic failure. Because of this, the mitigation of gouging is a primary concern. Figures 1.3 and 1.4 are of a total rail failure and catastrophic sled failure, respectively.

A typical gouge in the rail is shown in Fig. 1.5. This particular section of rail is coated with an iron-oxide paint developed to mitigate gouging. The shoe section also appears on the rail to further illustrate the geometry. A schematic of a typical gouge is presented in Fig. 1.6.

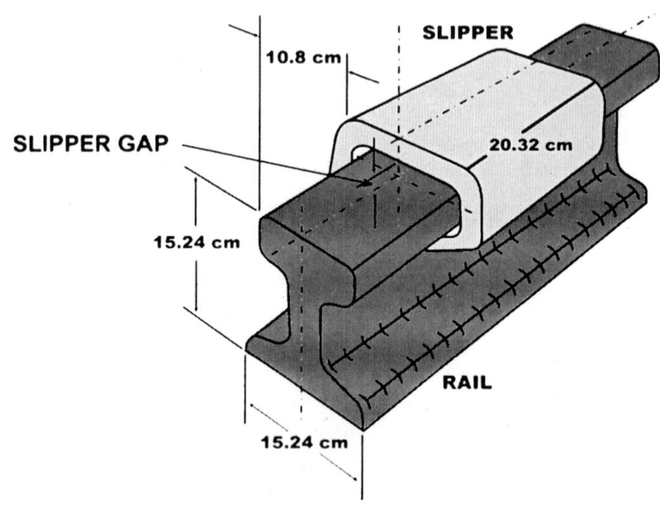

Fig. 1.2 Rocket sled shoe-rail interface.

Fig. 1.3 Typical total rail failure.

III. Gouging

To understand the initiation of the gouging phenomenon and hypervelocity speeds, and its mitigation, we will examine the following from both past and current research efforts in the field. Upon establishing this foundation, a detailed examination of the gouging phenomenon will be undertaken, with a metallurgical study of the gouge shown in Fig. 1.5. This study will indicate a verifiable thermodynamic history that will aid our understanding of the gouging event. To modify

Fig. 1.4 Catastrophic rocket sled failure.

Fig. 1.5 Typical rail and shoe gouge.

past modeling efforts to maximize accuracy, a study that determines the material constitutive models for VascoMax 300 and 1080 steel is presented. These material flow models are then validated using impact experiments and CTH simulations of the experiments. Additionally, a theoretical penetration model that indicates the impact energy necessary to initiate gouging, without the need for computationally intensive CTH simulations, is created.

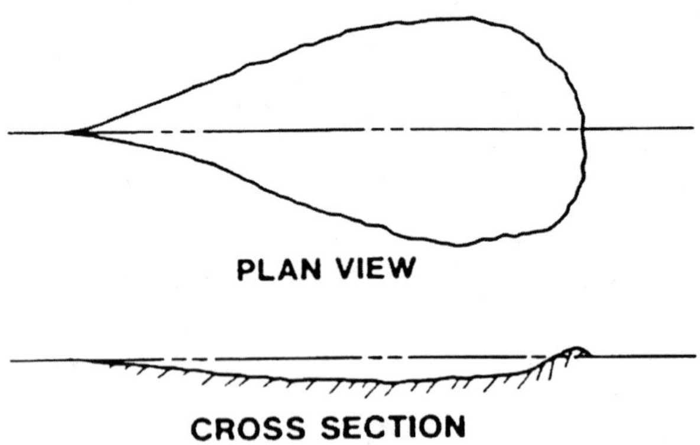

Fig. 1.6 Schematic of typical rail gouge.

Based on the extensive experimentation with the materials in the HHSTT gouging problem, a validated CTH model of the gouging scenario is conducted. These simulations suggest the mechanism for gouge initiation and design recommendations, which will improve the probability of success for the HHSTT's goal of achieving sled velocities on the order of Mach 10, are mode. A greater understanding of the initiation and mitigation of gouging is thereby attained (see also [1]).

Reference

[1] Cinnamon, J. D., "Analysis and Simulation of Hypervelocity Gouging Impacts AFIT/DS/ENY 06-01," Ph.D. Dissertation, Department of Aeronautics and Astronautics, Air Force Inst. of Technology, Wright-Patterson AFB, Dayton, OH, June 2006.

Chapter 2

Previous Research in the Hypervelocity Gouging Phenomenon

I. Introduction

HYPERVELOCITY gouging research has been an area of interest for many years. In fact, this author is following the fruitful work of Laird [1] and Szmerekovsky [2] under AFOSR sponsorship. This chapter will trace the previous research in this field by summarizing the work of Szmerekovsky [2] in the description of this history and providing a summary of his contribution to this research effort. To begin, a brief description of the gouging phenomenon is presented.

II. Description of Gouging

In this AFOSR effort to examine gouging, Laird [1] presented the definition of gouging as follows:

> Gouging is a failure mode found in metals undergoing hypervelocity sliding contact. When inertial forces are so great that the materials exhibit fluid like behavior, shock induced pressure creates a region of plasticity under the location of impact. Tangential motion of one body with respect to the other deforms or shears material at these points and results in deformation of the parallel surfaces that impinge on each other in a continuous interaction. Once this interaction region grows large enough to shear the surface of one of the materials from the bulk material, a gouge has been formed. Continuous interaction of the materials in the region of the gouge will cause the gouge to grow further until the materials are no longer in contact. [1]

This definition has several key aspects that delineate gouging from other types of material deformation. Laird found that the material interaction was characterized by the creating of material "jets." These jets contained thin plastic deformations at high strain rates. These jets would form from both sections of bulk material, and the event would proceed with the material jets mixing. In the HHSTT application, this gouging would involve material from the shoe and the rail. Recall that Chapter 1 described the catastrophic impact that gouging can have on the sled system. A single gouge can either be insignificant and thereby not cause sled failure and only require rail repair, or it can be large enough to be catastrophic to the test sled run. The prediction of the type of gouging or severity of the effect remains highly problematic. This interaction is depicted in an early numerical simulation result in Fig. 2.1.

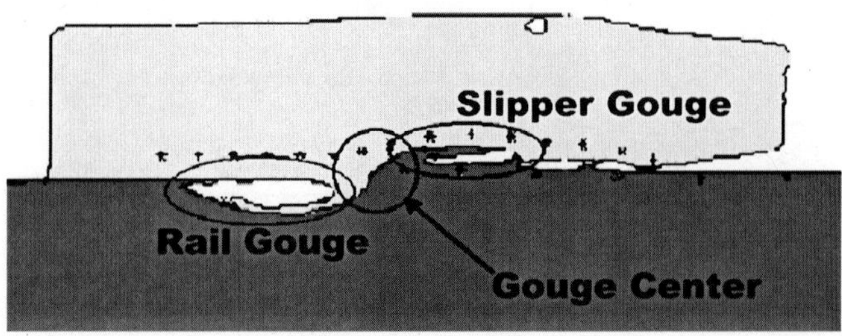

Fig. 2.1 Typical material gouging interaction.

Laird also presented a nominal sequence of events in gouge development. In numerical simulations, gouging was preceded by a hump of material being created plastically from the bulk material. This hump precipitated material flow in both the shoe and the rail, led to jetting, and eventual material mixing. This can be seen graphically in Fig. 2.2.

This type of material deformation does not occur at low velocities. The exact definition of hypervelocity and therefore the exact description of when this gouging phenomenon might occur is not a precise endeavor. However, as Laird reports, a widely accepted estimate is as the impact velocity approaches the order of magnitude of the elastic wave speed of the specific material gouging becomes more probable. This elastic wave speed c_0 is defined as

$$c_0 = \sqrt{\frac{E}{\rho}} \tag{2.1}$$

where E is the elastic modulus [which is the linear elastic slope of the equivalent uniaxial stress-strain (σ-ε) curve of the material] and ρ is the density of the material. Applying this relationship to the manufacturer reported values (Allvac [3] in the

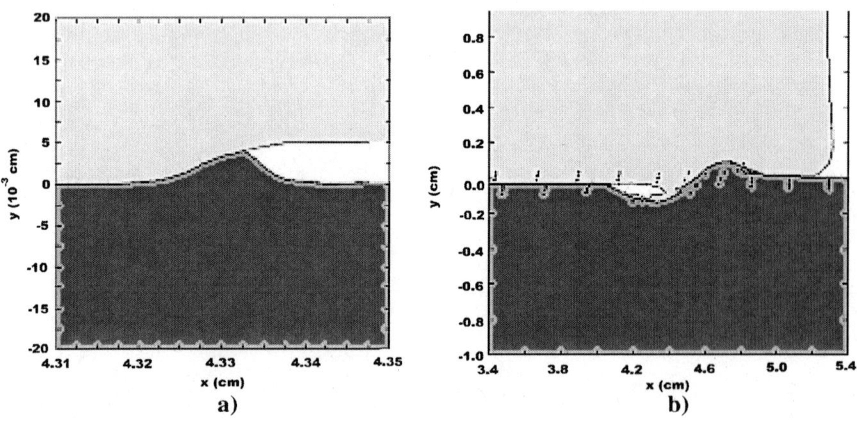

Fig. 2.2 Gouge a) initiation and b) development.

Table 2.1 Elastic wave speed for VascoMax 300 and 1080 steel

Material	E, GPa	ρ, g/cm^3	c_0, km/s
VascoMax 300	190	8.00	4.87
1080 Steel	205	7.85	5.11

case of VascoMax and U.S. Steel [4] for 1080) for both E and ρ for VascoMax 300 and 1080 steel results in Table 2.1.

Based on these calculations, we would expect gouging to start to occur as the impact speed becomes greater than 1 km/s. This is, in fact, the velocity range in which the problem begins and will be discussed in detail later in this chapter.

The high stress concentration present in these hypervelocity impacts leads us to modify Eq. (2.1) to account for the plasticity of the material in this manner:

$$c = \sqrt{\left(\frac{\partial \sigma}{\partial \varepsilon}\right) / \rho} \tag{2.2}$$

where c is the plastic wave speed and $\partial\sigma/\partial\varepsilon$ is the local slope of the stress-strain curve. Equation (2.2) becomes significant as the material deforms plastically and indicates that the creation and propagation of shock waves becomes significant [2].

Another depiction of the gouging event was proposed by Szmerekovsky [2] in Fig. 2.3 in which the following is noted:

1) Plastic displacement must create a steep amplitude above or below the datum sliding line.

2) For a gouge to develop, a relative velocity with respect to the bulk of the displacement must form at the portion of furthest penetration.

3) The portion of furthest penetration with a relative velocity is called the boundary layer portion of the plastic displacement.

4) The bulk of the displacement closest to the slide line is called the sublayer portion of the penetrating plastic displacement [2].

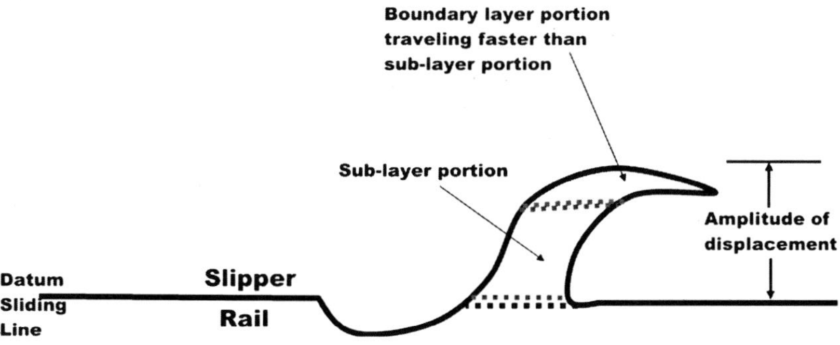

Fig. 2.3 Depiction of material jetting.

High-energy impact problems are therefore concerned with deformations that lead to the generation of both elastic and plastic waves. The plastic waves are created when the stresses present exceed the material yield strength. Beyond the elastic limit, the stress-strain behavior becomes very nonlinear and must be described by the strength model that accounts for this nonlinearity. The choice of constitutive model that describes this relationship becomes critical. Complicating the analysis further is the fact that the rapid deformations lead to high strain rates and possible shock-wave creation. Additionally, the description of a material failure criterion also is central to understanding how the impact event will unfold. As mentioned earlier, the thermodynamics of the deforming material must also be considered.

In a high-energy, -temperature, and strain-rate deformation, the equation of state (EOS) relationships of the material can dominate the solution. In the EOS, the pressure, density, and temperature relations of the material are defined. These relationships become necessary in the solution of the impact problem and provide the required additional equations that are coupled to the conservation equations to formulate the impact problem. The EOS relationships have the capability to model shock waves and fluid-like material behavior at high rates of deformation. They are central to numerical codes known as "hydrocodes" (see Zukas [5] and Anderson [6]).

As Laird [1] outlined in his definition of gouging, we are considering a high-energy impact event in which shock physics and nonclassical mechanics (such as nonlinear strength and failure models) dominate. It is within this context that the following summary [2] of research in the field of hypervelocity impact phenomena is presented.

III. Previous Hypervelocity Gouging Research

Previous research in this field can be delineated into the following six focus areas [2]: 1) test track observations and gouging tests, 2) laboratory gouging tests, 3) numerical modeling of gouging, 4) aerodynamic sled analysis, 5) load and failure analysis, and 6) methods for gouge mitigation.

A. Test Track Observations and Gouging Tests

One of the areas of examination in this field of hypervelocity gouging is the investigation of the test track runs and their resulting postgouge structures. The gouging phenomenon has been present at the HHSTT for over 50 years. The effort to understand the initiation of gouging and actions to mitigate it have been ongoing for that long as well.

In the late 1960s and early 1970s, Gerstle [7–9] examined this effect using a monorail test sled at the Sandia National Laboratory. One of his focus areas was to define the kind of rail characteristic that would initiate gouging. The thought, at that time, was that rail roughness (uneven rail height, either at a rail section joint or in the middle of a section) or debris on the rail (an asperity) would initiate the gouging process.

He found that gouges frequently occurred downstream from discontinuities in the rail surface (such as discrete rail section interfaces), but that three-dimensional, small radius (and thereby nonsharp), irregularities such as weld beads across the width of the rail did not cause gouging. This indicated that a discontinuity across the rail head would distribute the stress, cause uniform deformation, and prevent the development of a uniaxial strain condition. This condition would create a sharp wave front and would allow the formation of a shock wave (which is a plastic wave of uniaxial strain) [10]. The uniform deformation frustrates the shock-wave formation by relaxing the stress wave created by high speed impact. Without this shock wave, the high pressure gradients required for plastic flow and gouging are prevented.

Gerstle also performed a metallurgical study on the damaged rail section from his testing and discovered that the gouges had a surface layer of 304 stainless steel (sled shoe material used in his tests) deposited on top of martensitic 1080 steel (the rail material). The presence of martensite in the 1080 steel demonstrated that the temperatures were high enough to austenitize the steel, and then the steel rapidly quenched. He references several studies in [9] that indicate similar heating and quenching has been shown in punch tests. He also observed that the rail material was severely strained and microcracked. Gerstle believed this to be evidence of catastrophic thermoplastic shear (or adiabatic slip). Thermoplastic shear "occurs when the local rate of temperature change is such that the resulting strength decrease exceeds the rate of increase in strength due to effects of strain hardening" [9]. In other words, a large temperature change in a small localized area softens the material in that same area quicker than strain hardening strengthens it. It then becomes an area of local weakness in the material and thus a likely spot of shear fracture. During adiabatic slip in steel, for example, local heat generation is large enough to austenitize the material, but the large mass of metal around that thin shear zone of the austenite material will quench it quickly enough to turn it into martensite.

Gerstle also found shoe material (stainless steel) in the examined rail section. This indicated material mixing that is a characteristic of gouging. He also reported that the nongouged sections showed no damage, except for evidence of surface decarburization—indicated exposure to a high-temperature source (over 800 K typically).

Gerstle also examined the variations in cracking within the gouged region. He found evidence of adiabatic shear bands that formed normal and parallel to the direction of sled motion. This indicated that stress waves were interacting within the plastically deformed material. He found the most significant fracture on the gouge centerline (in the deepest section of the gouge). Additionally, he found martensite within these fracture areas, indicating a high thermal load, followed by rapid quenching.

In his investigations, Gerstle found that the high-energy impact event created a large thermodynamic event that allowed the material to experience phase change and rapid temperature variation. Additionally, the thermal energy imparted by the plasticity of the material might have been relieved by the creation of shear bands, which can be thought of as a thermal sink. Therefore, some of the energy in the event was absorbed by the material in this fashion.

Gerstle's work suggests that gouging is a thermodynamic event. The creation of shear bands is not necessarily an adiabatic process but can involve heat transfer within the material. Shear band formation, which leads to material failure, also creates an internal method of heat absorbtion within the impact event that needs to be investigated.

A summary of sled development and the associated problems with rail damage, gouging, and shoe wear was produced by Krupovage and Rasmussen [11] in 1982. They summarized research that has been done at Holloman AFB in various in-house reports. They identified gouging as being the result of impact between the shoe and the rail and that these impacts seemed to indicate a large bearing load. That is, the vibration of the sled as it travels down the rail results in a large compressive load on the rail as the sled comes down from an oscillation. They note that efforts to reduce this vibration by controlling the aerodynamic forces had been somewhat successful. They also describe the fact that the shoe/rail interaction is one of constant wearing contact, as evidenced by "slipper fire." This slipper fire is a constant stream of bright effluents from the back end of the slipper as it travels downrange. They attributed this intense light/fire to aerodynamic heating of the shoe and rail and the oxidation of the eroding shoe and rail material (i.e., from wear and not necessarily gouging).

The authors proposed the following relationship for the rate of work developed by friction that could create this frictional heat:

$$\dot{w} = C_f N v \tag{2.3}$$

where C_f is an empirical friction coefficient, N is the normal force, and v is the sled velocity. This equation assumes that the work is converted exclusively to frictional heat. Therefore, determination of the other parameters could provide an estimate of the energy acting as a heat input.

This extreme contact heat can cause the materials to melt. The authors, however, observed that melted metal acted more like an abrasive than as a lubricant. Because the concept of coating the rail to prevent gouging has been successfully implemented to some extent at the HHSTT, this discussion concerning the desirable properties of the coating becomes critical. If the liquid metals are abrasive and therefore increase the likelihood of damage, then the prevention of this melting is a goal of their effort.

In 1984, Krupovage [12] again examined rail gouging. He summarized numerous test sled runs with varying sled geometry, test conditions, and velocities. The largest gouge that he reported was one that measured 4 in. long, 3 in. wide, and 0.4 in. deep. Gouge location varied widely. Gouges were found at rail section interfaces and on the inside of the shoe surface. Some of the gouges contained shoe material and, in some cases, copper that originated from an aerodynamic wedge fixed to the front of the shoe.

Krupovage also noted that at high speed (above 5000 fps) sled material was shed from the forward section of the sleds. In these high-speed runs the aerodynamic heating effect was also observed. From these facts, Krupovage concludes that gouging is a function of the high aerodynamic heating load, liberation of shoe and rail materials from high heat and oxidation (i.e., wear products from shoe and rail), and debris from either the deterioration of the sled components or an outside source. He argues that gouging is initiated from this debris (referred to as an asperity in other references) interacting inside the gap between the shoe and the rail. Krupovage did not believe that simple (vibratory) impact would be sufficient to cause gouging in itself. An important distinction is that his definition of debris would also include rail section interfaces and rail surface roughness as asperities.

Krupovage also reported more gouging in the sled coast phase. (This is the phase in which the sled is no longer under propulsive acceleration.) On many of these test runs, the sled enters a large helium bag at the terminal end of the run (where the target of the sled payload resides). In this environment, he found no material loss from aerodynamic heating. From these observations, Krupovage proposes that a dynamic model consisting of the shoe undergoing vibratory impact with the rail should be constructed to model this gouging phenomenon.

Krupovage's work points to the importance of the aerodynamic heating to this vibratory-type impact problem. Additionally, his observations that gouging is initiated by an asperity of some kind is also key to describing this high-energy event.

Barber and Bauer [13] compared the phenomenon of sliding contact at low, high, and hypervelocity (velocity at which interaction forces are predominantly inertial) in 1982. (See also a summary of contact mechanics by Barber et al. [14] in 2000.) They argued that there existed a threshold velocity for hypervelocity gouging. Additionally, they developed a model for hypervelocity gouging that included the concept of impact with an asperity. Their description of gouging included the concept that two metals in sliding contact created "cold welds," which then sheared to create a thermal event. This thermal environment increases in intensity until the melt temperature of the metal is exceeded, and molten material are therefore created at the material interface. This molten layer is argued to reduce frictional effects and decrease gouging [13].

This description contradicts Krupovage and Rasmussen's [11] assertion that a liquid metal interface acts as an abrasive rather than to reduce interface friction. At hypervelocity, however, both investigations agree that frictional forces increase between a liquid and metal. Laird's description of gouging compares favorably with this (see Sec. II [1]). Laird's description of gouging centers on the development of a high pressure core generated from the plasticity of the impact. The plasticity of the material interacting with the motion of the shoe shears material from both the rail and the shoe, and this process initiates gouging.

Although this description by Barber and Bauer does consider the contribution of thermal loads/diffusion and shear stresses in the development of plasticity, it is somewhat incomplete in its description of gouging. The phenomenon of shear band formation, for instance, is not addressed. However, the emphasis on the heat-transfer aspects (i.e., the thermodynamics) points to the importance of this specific area of research.

Barber and Bauer define a "hypervelocity sliding threshold velocity" at which the stress created by impact exceeds the material ultimate strength. The impact stress, and thereby this threshold velocity, is described as being a function of impact velocity with an asperity, angle of impact, material density, and shock speed. They proposed that this impact created a very localized material defect (like a crater) that travels in the direction of the shoe at approximately $\frac{1}{2}$ the shoe's velocity. This would then result in the typical gouge shape. Although this mechanistic description offers to explain an asperity impact scenario, it does not address the simple vibratory impact case, known as an oblique impact. In this type of scenario, a gouge is created by the shoe impacting the rail at a shallow angle and creating a gouge without the necessity of an asperity. Additionally, Barber and Bauer's approach does not account for material mixing reported (e.g., in the metallurgical findings of Gerstle et al. [9]).

Barber and Bauer did not report significant correlation between their theories and quantitative data. They did, however, point to the work of Graff et al. [15], in which it was noted that there appeared to be a minimum shoe velocity and normal load required to initiate gouging. Therefore, a hypervelocity sliding threshold velocity was supported by the data. Barber and Bauer also noted that in some rail gun gouging studies (see Sec. III.B) a threshold velocity was discovered.

The concept of a minimum horizontal velocity required to initiate gouging is borne out in the experience of the HHSTT. However, many other variables that make such a simple generalization somewhat problematic enter into this analysis.

In 1997, Mixon [16] performed a survey of the literature and experimental data and created a detailed review of the gouging phenomenon. He examined a database of test runs and resulting gouging and summarized the common elements to describe the factors that contribute to gouge creation. These elements included high stress from dynamic loading, high sled velocity, asperities on the rail surface, heating caused by friction, deterioration of the shoe and creation of debris that lodges in the slipper gap, and ejection of sled material caused by aerodynamic heating (high temperature behind the normal shocks).

Mixon reviewed three types of tests for his analysis. They included the low mass interceptor (LMI), medium mass interceptor (MMI), and Patriot PAC3. Each of these test series consisted of discrete sled sections, with the payload being carried by the forebody sled and pushed by the pusher section. The shoes of the forebody, as well as the shoes of the final stage pusher rocket (called the Roadrunner), could initiate gouging.

The LMI tests (two of them were available) examined by Mixon made use of the terminal helium bag described earlier. The test runs had a peak velocity of 6863 fps and experienced gouging beginning at approximately 5800 fps. The majority of the gouges (75 to 83%) occurred past peak velocity, and all of them occurred in the helium environment. In this regime, the aerodynamic heating and oxidation is kept to a minimum, but the deterioration of the shoes has already occurred.

Seven MMI test runs were evaluated, with a peak velocity of 6660 fps. The location of the gouging on the railhead was recorded for four of these runs. The onset of gouging occurred at about 5400 fps, and a total of 408 gouges was discovered (with 24 being significant enough to require welding to repair). Figure 2.4 depicts the velocity profile (represented by the line) of the sled, as well as the location and number of the resulting gouges (shown as bars indicating the number of gouges per 500-ft section of rail). It is clear from Fig. 2.4 that most of the gouges are in the region of maximum velocity. The fact that the largest number of gouges occurs after peak velocity points to shoe degradation as a possible cause of the phenomenon. A majority of the gouges was found to occur at the rail corners, where stress and temperature concentrations are a maximum.

Mixon presented diagrams of the gouge locations and noted that there were sections of rail that had multiple gouges. Additionally, there were runs in which gouging was particulary prevalent. For instance, the worst run included 114 (27.9%) gouges, nine (37.5%) major gouges, and a large 6-in. gouge that broke the rail and led to catastrophic failure.

The Patriot (PAC3) tests included 14 runs with a peak velocity in the 6000–6100-fps range. This series of tests also included an aggressive rail repainting effort. The track was sandblasted and repainted with 6 mils (±1 mil) paint every

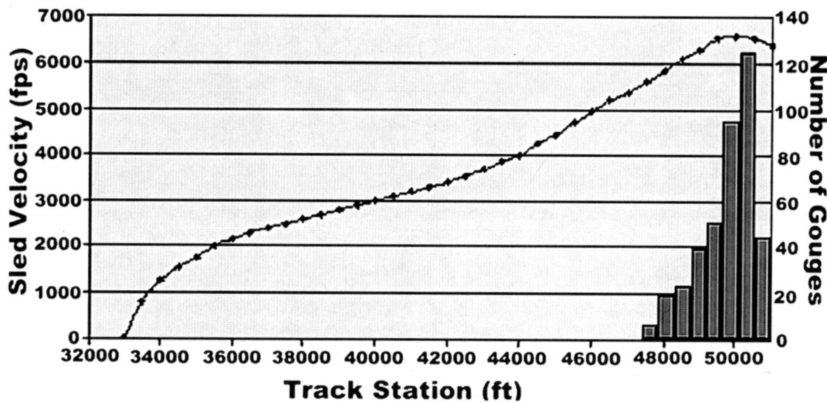

Fig. 2.4 Sled velocity vs number of gouges (MMI tests).

four test runs. The rail was also spot painted after every run, where needed. In this test series, gouging began at 5750 fps, and gouging seemed unrelated to whether the vehicle was before or after the peak velocity. All of the gouging was in the helium environment. This test series only suffered two major gouges (both on the same run). This same run ended with catastrophic failure of the Roadrunner.

Mixon argues from this analysis of the test record that gouging is related to the roll forces in the vehicle. He concludes that the location of the gouges shows that the sled experiences roll or lift during the run and that gouging leads to more excessive roll moments that contribute to total system failure. Mixon also notes that the fact that gouging occurs late in the test run indicates shoe wear/deterioration is a factor in gouge initiation. However, this region is also where velocity and aerodynamic heating is at a maximum, and so clear causality cannot be easily established. Additionally, the dynamic nature of propulsion cutoff (i.e., a sudden drag load imposed on the vehicle once the propulsion system turns off) is not addressed in the analysis. Mixon did find that the highest gouge rates were found in a small band of velocities ($\pm$50 fps) around the peak velocity.

Mixon concluded that high stress, high velocity, rail imperfections, deteriorated slipper surfaces, and frictional heating were the significant contributors to gouging. His summary pointed to the fact that rail coatings can reduce the onset of gouging. Therefore, as we will see in Sec. III.F, gouging mitigation efforts have focused on this area. In fact, Mixon asserts that an accurate gouging model is needed to study these rail coatings. However, he noted that the model needs to be capable of considering various coating material properties and to ascertain their effect on the gouging event.

In this section, analysis of gouged materials from hypervelocity test track runs has produced some theories on the causes and mechanisms of hypervelocity gouging. These include the formation of adiabatic shear bands and thermoplastic shear, high-temperature effects (both from aerodynamic heating and plasticity of the material), inertial effects of hypervelocity impact, and shock-wave formation. In addition, formation of a plastic zone, high strain rates, viscoshearing, and hydrodynamic bearing might also be mechanisms causing hypervelocity gouging based on analysis of gouged test track materials. Clearly there are numerous variables at play in this gouging phenomenon. From the research in this section, there appears to be a strong relationship between the velocity of the sled and the onset of gouging. These various observations and contributing factors point to the need for a robust and validated model to evaluate the phenomenon of hypervelocity gouging.

B. Laboratory Gouging Tests

In addition to the experimental testing available using test tracks as outlined in Sec. III.A, experimental data for hypervelocity gouging are available from laboratory gouging tests. In these tests, gouges are created under laboratory conditions by impacting a projectile into a target at high velocity. The materials are typically carefully studied in their "virgin" state and after the gouging has occurred.

As early as 1968, Graff and Dettloff [15] created a gun experiment that generated a high velocity sliding impact with velocities up to 9000 fps. In this effort, projectiles were shot at a shallow (grazing) angle against flat or curved plates (targets).

The authors began with a review of all available test sled data from the HHSTT. Their conclusions, based on the gouging information, were that factors such as rail and slipper materials, slipper geometry, rail roughness, airflow in the slipper gap, sled velocity, and contact stresses played a role in gouge development. They argued that gouging was the result of high-velocity sliding contact (impact) between two metallic bodies (surfaces). They noted that gouging seemed to initiate in the 5200- to 5550-fps range and that the typical gouge was tear shaped (2–4 in. long, 1 in. wide, 1/16 in. deep).

Graff and Dettloff noted that the gouges seemed to exhibit evidence of discoloration and material mixing (i.e., metal deposits within the gouge). They also observed that the largest number of gouges occurred after peak velocity was attained and that about 80% of the gouges were on the side or top edges of the rail, 15% were on the undersides, and only 5% were on the top surface of the rail. They argued that, although shoe deterioration plays a role, sled velocity was the primary factor in the gouging phenomenon. In addition, in those runs in which canards on the sled were set to push the sled down into the track (and thereby reduce vibration), there was less gouging. Also, the advent of high-strength maraging steel for the shoe material reduced gouging.

To create laboratory gouging, Graff and Dettloff focused on the variables of impact velocity, shoe and rail materials, and interface stresses. They created gouges in a steel target with projectiles fabricated from brass, copper, steel, and aluminum. They found that a 3-ft-radius curved steel target concentrated the stresses sufficiently to create gouges similar to those at the HHSTT (note that a 20-ft-radius plate did not).

Graff and Dettloff noted that the projectiles marked the target plate during the impact and that a layer of mixed projectile and target material was created. This indicated that gouging involved material mixing and that the temperature reached the melting point.

Graff and Dettloff also postulated that aerodynamic heating during a sled run could, in itself, raise the temperature of the shoe enough to cause material melting without the necessity of an impact. They argued that there was evidence that the coating material acts as a lubricant and thereby prevents the transmission of shearing forces to the rail. In this fashion, the cause for gouge initiation was avoided. They also proposed that melting products could, like the coating material, act as lubrication and prevent gouging. The rationale was that the liquid interface would only transmit the spherical (volume changing) stresses and not the deviatoric (shape changing) stresses. The only shear forces transmitted, then, would be within a narrow viscous boundary layer on the shoe.

Graff and Dettloff presented analytical computations which demonstrated that steady pressure dominated the impact event and that the transient stresses were relatively insignificant (based on projectile size and velocity). The magnitude of these normal stresses was, for a 0.27-oz steel projectile, around 78,000 psi.

They also demonstrated that gouges could be initiated by putting discontinuities on the target surface. These locations would experience stress concentrations and that would lead to gouging. Gouging could be increased by something as simple as orienting the surface finish transverse to the projectiles motion as compared to having the grain parallel to the velocity vector. It was because of this study that they concluded the surface preparation was an essential contributor to the gouging process.

Graff and Dettloff proposed that gouging occurred when metal-to-metal contact begins between the projectile and the target. This means that the impact must possess sufficient normal stress and impact velocity to allow the projectile to penetrate the oxide film on the target and the molten layer on either surface to create a metal on metal interface. They argue this contact would create a "weld" between the materials and precipitate gouging.

The penetration described by Graff and Dettloff could be initiated by a stress concentration resulting from the target surface condition. This would begin the mixing type process described in Sec. II by Laird in which material from both bodies begins to rotate about a high pressure core. The authors here proposed that the gouge would start very small, but would grow as the metal-to-metal contact persisted and shear of the material continued. The gouge would terminate with passage of the projectile back end beyond the affected area.

In 1970, Graff et al. [17] continued the study of gouging by examining various projectile and coating materials to make some generalizations concerning the onset of gouging. They found the following:

1) All metallic projectiles caused gouging.

2) Hard metals have a higher threshold gouging velocity.

3) Hard maraging steel gouged least, but excessive hardness could result in machining action.

4) A shoe with hardness slightly higher than the rail would give the best results.

5) Best coating results were from low-strength, low-density, nonmetallic materials.

6) Low-strength coatings allowed shear to occur in coating, not in target or projectile.

7) Gouging was prevented if metal-to-metal contact could be prevented.

The work by Graff et al. points to the importance of a coating to prevent the metal-to-metal contact that appears necessary to initiate the gouging event. The characteristics of the coatings and a more detailed discussion can be found in Sec. III.F.

In 1995 and 1997, Tarcza and Weldon [18, 19] studied this concept of a minimum threshold gouging velocity by shooting projectiles at a relatively low velocity. He proposed that gouging could occur a low velocity and that it was a function of material properties. His aim was to demonstrate this correlation to predict the onset of gouging. Tarcza created an experiment that related gouging to impact velocity and material strength. He also sought to demonstrate that there was a threshold gouging velocity, but that it was much lower than previously reported. Additionally, he wanted to design an experimental protocol that would be inexpensive and fairly easy to duplicate.

Tarcza's review of the relevant gouging literature, in particular those that related to the rocket sled and rail guns, resulted in a list of common theoretical contributions to the hypervelocity gouging phenomenon. All of these varied sources acknowledged the role of shoe velocity, stress between the contacting surfaces, and the material properties of the structures in the initiation of gouging.

Tarcza examined the past gouging data and proposed a linear relationship between the threshold gouging velocity and the specific yield strength (yield strength divided by density). This is depicted in Fig. 2.5. Extrapolating this relationship, he argued that a lead-on-lead impact scenario should allow gouging to occur at approximately 715 fps. A common thread throughout the literature that Tarcza found was that the gouging was characterized by the various different impact configurations as having a high pressure core that was created by the impact and continued to grow. This core was required to grow in size from the time of impact, or gouging would not commence.

Tarcza postulated that gouging was possible below the threshold gouging velocities reported in literature given a particular set of impact conditions and material properties. He argued that hypervelocity gouging occurred at velocities at which

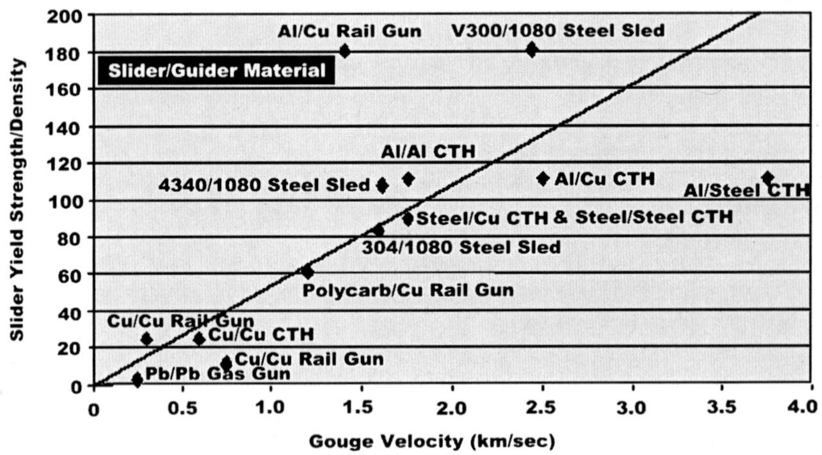

Fig. 2.5 Slider specific yield strength vs gouge velocity.

the inertial forces dominated the event and not necessarily at high relative striking speeds. He then sought a material combination that would make gouging possible in his laboratory environment.

Tarcza selected lead on lead as his material combination. He set up a light gas gun and shot .22-caliber projectiles at a curved target in a similar configuration to Graff and Dettloff [15]. He found that gouging occurred at striking (sliding) velocities of approximately 890 fps ($\approx$ 270 m/s). He found gouges in shapes similar to those reported in the literature for hypervelocity gouging events. Additionally, he found that gouging was related to the surface condition (i.e., irregularities, scratches, etc.), although not exclusively. Although these surface discontinuities contributed to gouging in most cases, other gouging events seemed simply the product of the material interaction. Counterintuitive to some thinking, a uniform seam across his lead target (much like a weld bead between rail sections) seemed to discourage gouging. Recall that this agrees with Gerstle's work in [7, 8].

Tarcza observed a form of deformation that lies between gouging and nongouging behavior. He called this "incipient" gouging. In this kind of interaction, gouges had been initiated, but had not fully developed. Additionally, as opposed to previous impact studies, Tarcza concluded that the presence of an oxide layer did not affect gouging. He argued that gouging was a result of sustained material contact and that the uniform target discontinuity (seam) prevented such sustained contact and thereby frustrated gouge development.

Although Tarcza was able to create gouging far below the widely held definition of hypervelocity (i.e., where impact velocity is close to the material sound speed), it does not invalidate aspects such as shock waves as contributors to gouging. It does demonstrate, however, that material properties and impact conditions play a significant role in this study also.

Tarcza also concluded that the normal force level was a crucial component to the development of gouging. This normal force is generated by an oblique impact angle with the target or impact with a surface asperity. Graff and Dettloff [15] and Tarcza [18, 19] both found that a curved target plate was necessary to initiate this gouging, i.e., that sufficient normal force was required. The normal force is assumed to be a function of the projectile (sled/shoe) velocity, but no data on the normal force generated are available. Therefore, assumptions with regard to the dynamic environment, amplitudes of oscillation, and sled orientation must be made to compute an estimate for the normal force of impact.

Ramjaun et al. [20] investigated the field of hypervelocity impacts with regard to space debris bumper shields in 2000. They examined hypervelocity impact craters formed at various striking angles at around 5 km/s. Their research discussed failure mechanisms that can apply to the shoe/rail interaction at very similar velocities. They discovered adiabatic shear band formation in the impact damaged area.

Ramjaun et al. [20] detail a failure mechanism in which the compressive shock wave of impact propagates within the projectile and the target (in our case, the shoe and the rail) and reflects off of the free surface. These wave reflections leave the material in a highly energetic state that can lead to fragmentation, melting, or vaporization.

The oblique impact scenarios performed by these authors are the most applicable to our gouging problem. Examination of the impact craters from the oblique angle shots reveals a gouge-like shape and characteristic material melting, mixture, and thermoplastic shear. Metallurgical evaluation determined the existence of adiabatic shear bands and intense plastic deformation. The authors concluded a large temperature excursion was experienced by the material.

Ramjaun et al. [20] concluded that the shear bands were the result of shear instabilities created under high stress and strain rates. This plasticity generates heat that concentrates in these shear band areas and causes thermal softening. When the shock-wave reflections return to the area, tensile stresses that fracture the material along these lines of shear instability are produced. These shear bands can link up and cause sections to break from the bulk material. Additionally, this seems to indicate a potential thermal sink mechanism as the thermal energy concentrates here and then is released.

These authors cited the primary cause of hypervelocity impact damage to be the formation of adiabatic shear bands. They conclude that to mitigate damage, the material chosen should "have no tendency to form adiabatic shears" [20]. The material to best resist hypervelocity impact damage would 1) have uniform and homogenous flow properties during viscoplastic deformation to prevent formation of adiabatic shear bands caused by uneven formation of viscoplastic zones; 2) have a high melting point to prevent cracking in case of adiabatic shear band formation; and 3) not transform into a brittle phase during shock loading, which increases the likelihood of fracture under loading.

These authors, then, describe specific material properties that should be kept in mind as we evaluate shoe/rail designs to mitigate hypervelocity gouging. This is applicable in our problem because, as Gerstle reported and as we will see in Sec. III of Chapter 4, shear band formation is prevalent in this gouging phenomenon.

These laboratory experiments in the field of gouging are extremely valuable and, unfortunately, rare. This is because of the high cost of performing such tests. To gain better insight into the phenomenon and because of the high cost of creating and running such tests, numerical investigation of gouging has taken place in parallel with experimental procedures such as test track observations and laboratory testing. These numerical simulations can offer insight into the creation of gouging and challenge us to recreate the experimental results within a model.

C. Numerical Modeling of Gouging

Because of the high cost of experimental work in the gouging field and the desire to create models for predicting/mitigating hypervelocity gouging, numerical models or simulations form an important basis of the field of research. These numerical approaches are typically based on, and validated by, the experiment work described in Secs. III.A and B. Numerical models can offer valuable insights into the physical understanding of the impact event. They also allow theoretical considerations to be tested without the expense of actual experimentation. Of course, it must be recognized that the models inherit the inaccuracies of our limited understanding of all of the physical processes involved in hypervelocity gouging.

Traditional computational methods in structural mechanics are based upon Hamiltonian mechanics in which the forcing function $F(t)$ is known. Thus, the system of equations based on Hamilton's equations of motion can be represented by

$$F(t) = [M]\{\ddot{u}\} + [C]\{\dot{u}\} + [k_{\text{nonlinear}} + k_{\text{linear}}]\{u\} \tag{2.4}$$

where $[M]$ is the mass matrix, $[C]$ is the viscous damping matrix, $[k_{\text{linear}} + k_{\text{nonlinear}}]$ is the stiffness matrix containing both linear and nonlinear terms, $\{\ddot{u}\}$ is the acceleration vector, $\{\dot{u}\}$ is the velocity vector, $\{u\}$ is the displacement vector, and $F(t)$ is the forcing function (also the vector of the applied forces). This equation is the result of solving the three fundamental conservation equations (a detailed explanation of which appears in Chapter 3).

These Hamiltonian dynamics are considered under traditional (nonshock) load conditions. Therefore, numerous aspects of a hypervelocity impact are not accounted for. For instance, material failure, viscoplasticity, large material movement, shock waves, thermal effects, and the like are not modeled. The forcing function $F(t)$ in Eq. (2.4) is assumed to be applied in rates below the wave speed of the material. However, we know this is not the case in hypervelocity impact problems. Inertial effects are more dominant in the solution of the fundamental laws of conservation and equilibrium. Additionally, at hypervelocity, high pressure can cause the materials under consideration to behave as inviscid fluids [10]. At these impact pressures, the equation of state begins to dominate the solution of the material deformation (see Chapter 3), which can account for the nonequilibrium thermodynamics and related effects that arise from this high-energy event.

Therefore, the numerical modeling of hypervelocity impacts must take into account this nonclassical material behavior and consider rapid loading and material response. As mentioned earlier, the high-energy impact event requires consideration of the EOS of the materials, as well as the constitutive modeling.

Boehman et al. [21] in 1977 published the earliest computational hypervelocity gouging model. Their computer model attempted to capture friction, wear, and gouging between the shoe and the rail. Their efforts to establish gouging criteria were largely unsuccessful, yet they were able to identify velocity regimes that were more stable.

Barker et al. [22] continued this numerical work in 1987 at the Sandia National Laboratory using the predecessor to CTH, CSQ. A parallel impact thermodynamics (PIT) model was created to model the shoe/rail impact problem. The name indicated the nature of the impact (parallel to the rail) and the use of a thermodynamics (i.e., a hydrocode) solution of the gouging event. Barker utilized an elastic/perfectly plastic constitutive formulation that generated a viscoplastic response beyond the yield strength. The gouging was initiated in this model by the use of an asperity on the rail surface. Figure 2.6 shows this model and the small gap between the shoe and the rail.

Barker et al. [22] believed the gouging phenomenon to be an impact-related event and created the PIT model and theory proceeding from that assumption. The hydrocode solved this impact problem considering shock-wave physics and thermodynamics. To quantify the frictional heat generated, they relied on experimental work involving a 30-mm-diam steel ball shot down a barrel (with a curvature of 1 mil per 10 in.) at 3 km/s. This resulted in the surface of the ball melting after 2000 μs (60 cm of travel) to a depth of 6.7 mm.

Barker et al.'s model was validated in that it produced gouging when the shoe impacted an asperity at high velocity. The leading edge of the shoe was necessarily created at a 45-deg angle to crush the asperity under the shoe during the impact sequence. In addition to this two-dimensional model, they created a three-dimensional model in which gouging was shown to occur, but only if the gap

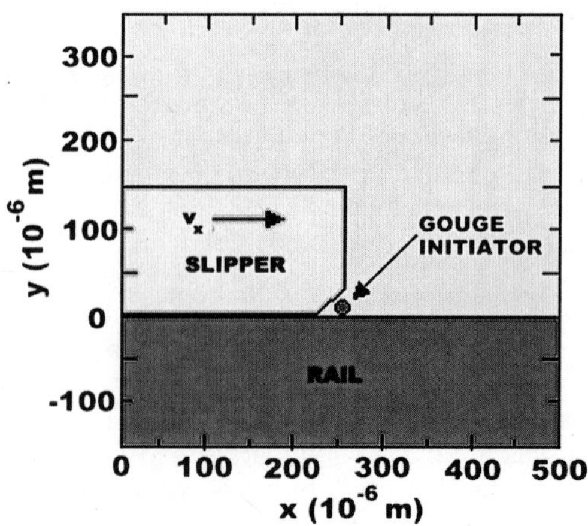

Fig. 2.6 Barker's PIT model for gouging.

between the shoe and the rail were removed. Because gouging was successfully recreated in the code, they authors felt the PIT model was validated.

Their study of the PIT model showed that the gouging was characterized by extreme local deformation (of the asperity in particular), high heating and subsequent melting/vaporization of the materials, and the creation of a high-pressure region. They theorized that this high-pressure interaction would deform the otherwise parallel material interfaces in such a manner that they impinged on each other and began to mix, thus producing a gouge. They postulated that once this interaction began it could become self-sustaining and would continue until the back end of the shoe past the interacting region. (This matches the description proposed from the experimental work in Secs. III.A and B.) They also noted the importance of the shock interaction within the gouging event.

Barker et al. performed a parametric study on his model results to better quantify the contributions to gouging from various variables and to offer suggestions on gouge mitigation. This study also served to verify the validity of the PIT model assumptions. The study concluded that in order to mitigate gouging one needed to 1) increase the gap size between the shoe and rail, 2) increase the shoe yield strength with respect to the rail, 3) use plastic as the shoe material, 4) pitch the shoe to create a small angle between the shoe and rail, and 5) decrease the normal load between the shoe and rail. Interestingly, the effect of the normal force in the creation of gouging was recognized as a key component.

Barker et al. ran his model without friction to test the hypothesis concerning gouging proposed by Graff and Dettloff [15]. Recall that this theory was that gouging was the outgrowth of metal-on-metal contact resulting from the projectile penetrating the oxide layer and the layer of melted material between the projectile and the target. In Barker et al.'s model, gouging occurred with or without friction. The friction was removed by considering a layer of frictionless material between the shoe and the rail. Therefore, the presence or absence of the frictionless material between the projectile and the target did not significantly affect gouge initiation.

This indicated that the inertial forces dominated the impact event rather than the formation of a welded junction as proposed by Graff and Dettloff.

With this analysis in place, Barker designed a laminated shoe that specifically allowed release waves to travel faster and thereby relieve the pressure in the high pressure core. This served to decrease peak normal pressure and also allowed for melt lubrication at high velocity. This design was fielded and reached 1.9 km/s without gouging in testing.

Although the PIT model offered insight into the gouging event, there were several shortcomings. The shoe design in the model was not accurate, and the actual shoe (with a much less steep angle of attack to the track) still was thought to lead to gouging where that geometry in the model would not. Additionally, an asperity was required for gouge initiation where it was thought that gouging was possible without one.

Another area where hypervelocity gouging occurs is in the development of rail gun technology or high-speed multiple-stage gas guns. In these applications, gouging can occur during the projectiles movement down the gun barrel, resulting in very undesirable outcomes. Barker et al. in 1989 [23] reviewed Susoeff and Hawke's 1988 report [24] on rail gun gouging. Barker et al. concluded that, although the source of gouging damage was not certain, it was possible that the projectile had shed molten material which had impinged into the barrel and precipitated the damage. The lack of gouging at very high velocity could be caused by the projectile completely vaporizing, whereas at lower shot velocities the projectile survives to damage the barrel.

Barker et al. also presented another parametric study of CTH conditions that lead to gouging within the PIT model. The materials of copper, steel, aluminum, and plastic were evaluated in all combinations and in the velocity range of $\frac{1}{2}$ to 12 km/s. The aim of the study was to describe whether an asperity would result in a growing interaction (pressure) region and create gouging or not. Barker concluded that there was a minimum and maximum velocity range in which gouging would occur. That is, the already discussed threshold gouging velocity also had an upper limit according to the model. (This upper limit has not been demonstrated in experimentation.) The upper limit occurred when the impact velocity exceeds twice the wave velocities of the materials. Apparently the material has insufficient time to be moved into the interacting region, and therefore the high pressure core fails to grow. Additionally, the lower threshold velocity could be raised by increasing the material yield strengths.

Again the critical nature of material yield strength, normal force, shock physics, and material interaction was noted in the investigation into hypervelocity gouging.

Tachau continued the effort to model hypervelocity gouging in 1991 [25]. He began with a summary of the literature in the field and identified Barker et al.'s model as a starting point. He noted that the PIT model required an asperity, a gap between the shoe and the rail, and a downward crushing of the asperity to initiate gouging. He also noted that the PIT model neglected the effects of sliding friction.

Tachau argued that gouging could be initiated by an oblique impact and that a gap and an asperity were not required elements. Tachau created this model in CTH and improved on the PIT model by giving the shoe an initial vertical (downward)

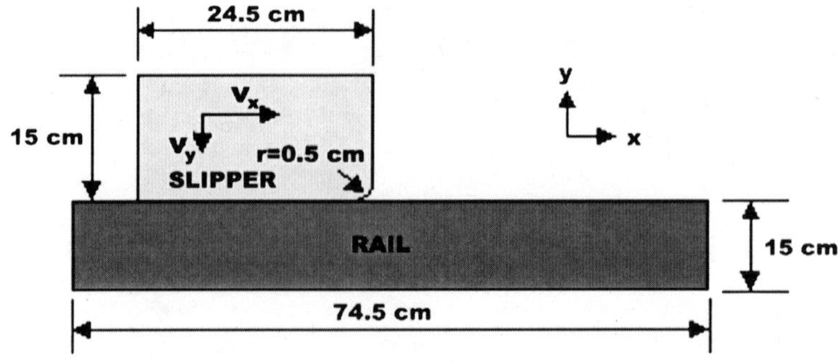

Fig. 2.7 Tachau's model for gouging.

velocity into the rail surface (see Fig. 2.7). Given the high horizontal (downrange) velocity of the shoe, this model would create highly oblique (very shallow angles) impacts. Tachau postulated that this would create the antisymmetric humps as described by Abrahamson and Goodier in 1961 [26] and would begin the gouging event.

Tachau's initial conditions were 2 km/s horizontally and 100 m/s vertically. [At this point Tachau's assumption of a 100-m/s downward vector was significantly high. In 2000, Hooser [27] (also, Hooser, Michael, "Dynamic Design and Analysis System Simulations," unpublished data, Holloman AFB, NM, 2001), using the Dynamic Analysis and Design System (DADS), showed that a more realistic vertical impact velocity is approximately 1 to 2 m/s.] Tachau observed that the resulting crater depths were deeper than those created in the actual sled tests. Additionally, his model showed very high temperatures (1800 K) resulting from plastic deformation near the material interface. This thermal input heated the surface to the melting point. Also, the core pressure generated in the resulting gouge region was around 5 GPa.

Tachau examined the model output for aluminum and steel shoes and varying velocities. He found that a high-pressure core is characteristic of gouging and that the initial velocities played a major role in whether or not gouging occurred. For example, for steel on steel, the 2-km/s horizontal and 100-m/s vertical velocities did cause gouging, but reducing the horizontal velocity to 1 km/s did not.

The results of Tachau's model led him to surmise that the high temperatures generated at the contact surface are sufficiently high to cause the interacting materials to thermally soften and flow. This interacting region of plastic material then allows the formation of a high-pressure core. The source of this high temperature, he argues, is from frictional effects and the energy of impact. These would be higher with increased impact velocities and normal loads.

For the oblique impact, this zone of viscoplastic material allows deformation and creation of the hump of material already discussed. This would then lead to material mixing and growth of the high-pressure core. This would be similar in nature to the gouging generated by the PIT model. For the impact with an asperity, Tachau demonstrated then as long as the asperity had a sloped surface, the impact would impart sufficient downward velocity that gouging would commence. This indicates that the vertical velocity of the shoe does not need to be significant in order to create a gouge in this model.

Once more, the topics of surface condition, frictional effects, normal force, and vertical impact velocity are the factors that alter the probability of gouging within this particular model.

Tachau extended his study in 1994–1995 [28, 29] along a similar vein. He concluded that a series of CTH studies showed that the gouging phenomenon was initiated by an oblique impact that causes a sharp temperature rise. This high temperature thermally softened the target and caused the characteristic hump of material to form the precedes gouging. The high-pressure core formed next, and the gouge was thereby created. In his work, Tachau relied on an elastic perfectly plastic constitutive law because a more complex relationship was not available in CTH at that time.

Following up on Tachau's work in 1998, Schmitz et al. [30] developed another model based on CTH results to examine gouging and wear. This particular model proposed to predict shoe wear and the onset of gouging based on empirical data and set initial conditions. Schmitz had the expectation that additional experimental testing would be performed to validate his model. In the creation of his model, he utilized an asperity impact simulation (see Fig. 2.8) based on Barker's model as described in [22]. Schmitz found that gouging was dependent on the creation of a growing high-pressure core within the first 4 μs of impact. He was able to correlate gouging in varying material combinations to experimental HHSTT data, which appears in Fig. 2.9.

In 2002, Laird [1] extended this investigation of hypervelocity gouging with an emphasis on understanding the fundamental physics of the phenomenon. Laird's focus was on understanding gouge initiation (in terms of material jetting) and the effect of temperature on the resulting gouging. In this undertaking, he performed a numerical study of gouging using the CTH hydrocode [31]. He also investigated

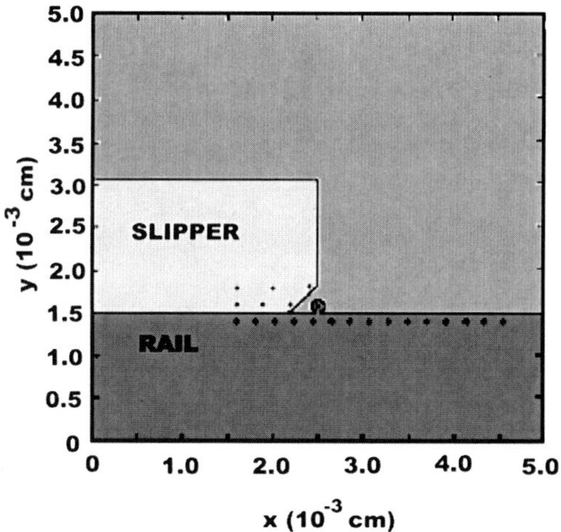

Fig. 2.8 Schmitz's model for gouging.

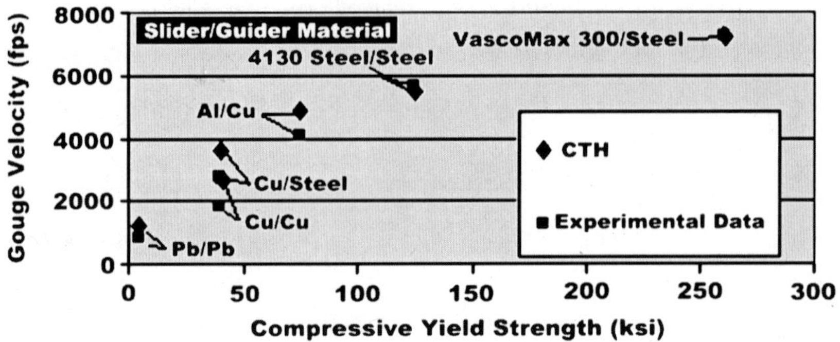

Fig. 2.9 Schmitz's validation of CTH to experimental data.

the effect that high temperature had on the resulting gouge [32]. Laird's model was created after scaling down Tachau's model by an order of magnitude. Although his work offers a comprehensive examination of the factors involved in gouging, his model was not scaled down using the Buckingham pi approach or similar mathematical scaling law [2].

The key elements of gouging identified by Laird include the plastic deformation of the materials, their strength, and the normal force. These factors must appear in combination (i.e., one alone is not sufficient) to initiate a gouge. The essential feature was a material jet (see Fig. 2.3) in both the shoe and the rail that began a material interaction that led to material mixture and eventual gouging. The jets were characterized by viscoplasticity of both materials.

He also argued that when these jets begin to form and initiate the gouging process, the reflected shock waves had not had sufficient time to return as tensile waves, and therefore a spall or tensile fracture is unlikely. He did note, however, that the high compressive stresses created an environment conducive to gouge development.

Laird also performed numerical examination of the high temperatures involved in the gouging event. He argued that these temperatures caused thermal softening and thereby reduced resistance to gouging. Indeed, whereas a room temperature impact would lead to gouging, one in which the shoe had been preheated (by aerodynamic heating in front of the sled, for instance) had a "jump start" to these higher temperatures and thereby making gouging more likely. The major difference between these two cases was that the preheated shoe created gouging earlier in time than the room-temperature case.

Additionally, Laird found that a redesigned shoe leading edge with a very shallow angle (less than 1.790 deg) did not gouge under the same conditions that a shoe with a rounded leading edge, caused by a shallow slope at the material interface, did. Therefore, the geometry of the shoe could, in itself, prevent the formation of material jets and therefore gouging.

Laird also discovered that increasing the material yield strength of the rail would also inhibit gouging to some extent. This analysis fit with previous work that showed that each of these parameters played a role in where the threshold gouging speed would be for a given geometry. Although this increased rail yield strength did not decrease the total penetration of the shoe into the rail, it decreased the viscoplastic interaction, and therefore the high-pressure core, the leads to gouge formation.

In 2004, Szmerekovsky and colleagues [2, 33–36] extended this research area by creating a CTH model of the impact using actual test sled dimensions. The details of his investigation will be discussed in Sec. IV.

Although Szmerekovsky addressed the previous limitations in modeling by creating a numerical model based on the actual test sled dimensions, several other limitations are inherent in these past simulations. Primary among them is that CTH models which most accurately model the hypervelocity impact event do not contain material property values or strength models specific to the materials used at the HHSTT. As a complement to this numerical study, another focus area for research has been on the aerodynamic effects of these hypervelocity speeds on the gouging problem.

D. Aerodynamic Sled Analysis

As we have seen in the preceding sections, the thermodynamics of impact plays a significant role in the gouging phenomenon. As some of the researchers have noted in previous years, the thermal environment is not only limited to the heat generated from plastic material flow, but originates also from the aerodynamics of the sled traveling down the rail at speeds of Mach 5 and higher. The high-speed passage of the sled through the air creates strong shock fronts that raise the stagnation temperature behind the shock and flows heat into the shoe/rail system. Based on the time scale required for heat conduction, however, the shoe is the only element of the interaction that will experience significant temperature effects.

As early as 1968, Korkegi and Briggs [37, 38] developed a two-dimensional analytical model to study the shoe/rail gap region under this high flow condition. They divided the flow region into four discrete areas: 1) a laminar flow near the stagnation point at the front of the slipper, 2) a turbulent boundary-layer region before the upper and lower boundary layers merge, 3) a merged region, and 4) a Couette flow asymptote (flow between a moving plate and parallel stationary plate).

By performing this analysis, they found that the air that flows through the gap is compressed by the shock front to significantly high temperatures and pressures. This results in high lift loads and heat gradients along the inner surface of the shoe. In the speed range we are examining (Mach 5 to 10), that confined flow in the gap reaches temperatures equivalent to those on the leading-edge stagnation points and also on the same order as those generated by sliding friction. For example, at 10,000 fps, the heating rates were on the order of 100 Btu/ft^2-s, which was reported to be close to that of a frictional heating when the shoe and rail are in sliding contact. This indicates that extreme heating will be present in this analysis, whether the shoe is in contact with the rail or not. Based on the conduction analysis presented in Chapter 4, however, there is insufficient time for this generated heat to conduct into the rail. The shoe, on the other hand, can heat over the duration of a test run and can experience elevated temperatures. An examination of the effect of a heat shoe on a hypervelocity impact is examined in Chaper 9.

Korkegi and Briggs developed an expression for gap pressure p as a function of the distance from the slipper leading edge x (see Fig. 2.10) from one-dimensional

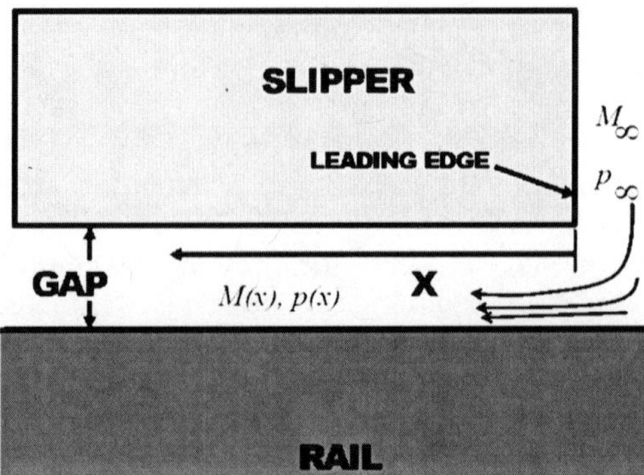

Fig. 2.10 Dimensional model for Korkegi and Briggs equation.

isentropic flow relations relating effective area to local Mach number and pressure as follows:

$$\frac{p(x)}{p_\infty} = \left[\frac{(\gamma+1)M_\infty^2}{2+(\gamma-1)M^2(x)}\right]^{\frac{\gamma}{\gamma-1}} \left[\frac{\gamma+1}{2\gamma M_\infty^2-(\gamma-1)}\right]^{\frac{\gamma}{\gamma-1}} \tag{2.5}$$

where p_∞ is the pressure at the freestream, $M(x)$ is the Mach number as a function of the downstream distance from the slipper leading edge x, M_∞ is the Mach number at the freestream, and γ is the ratio of the specific heats. The model is valid from the leading edge of the slipper to location where the upper and lower boundary layers meet. The model was developed for $M \gg 1$ and should therefore be valid in the velocity regime being studied. This equation is also valid for the helium environment when the proper Mach number and value of γ is applied.

Korkegi and Briggs concluded that the flow conditions varied dramatically between a cold shoe and a heated one. After the shoe heats up, which happens fairly quickly, a state of constant pressure exists between the shoe and the rail. However, a state of dynamic instability results with regard to pressure and the gap between the shoe and the rail. They showed that, as the gap narrowed, the pressure dropped off. Conversely, as the gap widened, the pressure increased. This instability results in the bouncing of the shoe against the rail and sets up the vibratory impact scenario observed in test runs. Therefore, not only does heat play a major role in the gouging event, but the aerodynamics also set up a dynamically unstable system that creates the environment for the vertical impacts and high normal stresses highlighted by other authors as key mechanisms for gouge initiation. This instability leads to the vertical impact velocities used in Chapter 9 in the modeling effort.

An external flow analysis was performed by Lofthouse et al. [39] in 2002 on the sled currently being used at the HHSTT. His computations were limited to an inviscid computational-fluid-dynamics solution over a velocity range of Mach 2 to 5. He discovered the highest pressure gradients generated by the sled occurred on the outer shoe surfaces. He considered the flow through the shoe/rail gap and predicted the pressure between the shoe and the rail to be characterized by shock interactions and sharp rises in pressure (jumps up to 75 psi). These pressure differentials could drive large temperature flows on the slipper. Again, this solution was inviscid. The addition of viscous flow could increase this effect significantly.

In the consideration of the aerodynamic effects on the gouging phenomenon, the literature shows a large heat flow in the problem. This additional source of heat adds to that already being generated by the plasticity and magnifies the thermodynamic contribution to gouge creation. Additionally, the aerodynamics have been shown to induce instability that causes the oscillatory impacts observed as one of the mechanisms responsible for gouging.

E. Load and Failure Analysis

Added to the numerous considerations just presented in the high-energy impact environment is one of material loading and subsequent failure. The high pressures and loads experienced by the structures necessitate a good description of the material failure mechanisms. The creation of shear bands and material jetting and/or mixing demonstrates the existence of material failure in this gouging phenomenon. Failure and damage research is focused primarily on developing the theory used for setting criteria for material failure, including thermodynamics of deformation and damage.

In 1961, Abrahamson and Goodier [26] noted that moving loads on soft or viscous materials were often preceded by "humps" of material. These humps were argued to be the result of inelastic material behavior. They postulated that if the material were elastic, the deformation would create equal humps before and after the moving load. If the material is moving relative to the load, the leading hump is drawn under the load. The resulting profile is a function, then, of the penetration into the target material and the horizontal velocity of the load. This characteristic hump is a key feature of the gouging process.

Voyiadjis et al. [40–43] and Abu Al-Rub and Voyiadjis [44] have developed a framework for analysis of heterogenous media that assessed a strong coupling between viscoplasticity and anisotropic viscodamage evolution for impact problems using thermodynamic laws and nonlinear continuum mechanics. Their proposed development included thermoelastic viscoplasticity with anisotropic thermoelastic viscodamage, a dynamic yield criterion of a von Mises type and a dynamic viscodamage criterion, the associated flow rules, nonlinear strain hardening, strain rate hardening, and thermal softening. The model presented in the research offers to be considered as a framework to derive various nonlocal and gradient viscoplasticity and viscodamage theories by introducing simplifying assumptions.

This theoretical development of a framework for a damage model is an example of development of a thermodynamic damage and failure model that could be used to improve the definition of failure for high-velocity problems such as hypervelocity gouging. Subsequent use of this model could aid in the understanding of the failure mechanisms involved in gouging. The primary limitation of this approach is the integration of this analysis into CTH or other shock physics codes. Although this failure model has been added to a Lagrangian finite element type code, its inclusion in an Eulerian shock physics code seems unlikely in the near term. This essentially makes this approach have no application to this type of analysis because all meaningful solutions use shock codes. Additionally, because it is not available in a shock-wave code, there is no linkage to an equation-of-state model, which could account for the nonequilibrium thermodynamic phenomenon.

In a related vein of research, Hanagud [45–47] is currently investigating a set of constitutive equations for high-energy impact under a state of nonequilibrium thermodynamics. The objectives of this research follow:

1) Formulate constitutive models and equations of conservation, for metallic projectile materials, in appropriate continuum mechanics and nonequilibrium thermodynamics framework. The formulated models should be able to explain shock-induced phase changes (including melting).

2) Simplify the constitutive model, as found necessary, and use the model, with other equations of conservation and interface conditions, to understand the penetration mechanism of metallic projectiles into isotropic and granular media at high initial impact velocities (e.g., 850 to 2000 m/s). The term understanding the penetration mechanism includes the projectile phase changes, melting, any failure of the projectile, and deviation of the trajectory from the intended trajectory.

3) Determine the parameters of the constitutive model and the penetration mechanism through testing.

4) Design new materials, their microstructure, and the spatial variation of the thermomechanical characteristics and structural design of the projectile to avoid trajectory deviation and any failure of the projectile.

The Hanagud constitutive models can be used to better describe the thermoplastic failure mechanisms of gouging. To accurately describe phase transition and nonequilibrium thermodynamics in which the first and second laws of thermodynamics are of uttermost importance, the Hanagud constitutive model is required. Most constitutive models assume adiabatic or isothermal states of thermodynamics. A similar limitation in application also applied to this work as had been applied to Voyiadjis in that inclusion of these developing theories into usable code is a problematic process. In addition, most of the nonequilibrium thermodynamic effects are at pressures and strain rates where the equation of state dominates the solution, not the constitutive model.

Central to this discussion is the recognition that in the areas deforming at high pressure and temperature the EOS relationships tend to dominate the solution, which incorporate nonequilibrium thermodynamic characteristics. This would lead us to focus on shock physics codes and analysis to describe the material interaction. However, the material flow model contributes significantly to the solution of the remaining areas that are deforming at lower pressures and temperatures [48]. Therefore, a poor constitutive model for the material could render the code simulations of the entire problem useless.

This area of performing load and failure analysis is central to the effort to understand and mitigate gouging. Current gouge mitigation can be summarized in the following section.

F. Methods for Gouge Mitigation

As described in the preceding sections, the high energy of impact can be absorbed by the materials in the form of a damage mechanism. Shear bands are one such mechanism examined already. Current gouge mitigation efforts revolve around coating the materials to improve their resistance to gouging. A coating and/or a change in material hardness/strength might be used to improve the material's resistance to impact deformation. In addition, the coating can prevent the transmission of shearing loads to the underlying bulk material or can change the thermal resistance of the material to the extreme thermal loads of impact. The thermal cycling of the coating can also affect its properties. Therefore, the area of coatings that can mitigate gouging by altering the impact environment is one of intense interest.

The HHSTT has successfully used coatings to reduce the occurrence of gouging. The analysis presented later in this section suggests that this is caused by the reduction of frictional effects (heat), the sacrificial nature of the coating to disallow shear stress from transmitting through it, the property of coatings to mask rail roughness, and its ability to protect the materials from the high thermal loads present in impact.

Coatings fall into one of two categories. The first is refractory. These coatings typically protect sled components from the harsh thermodynamic conditions, but can also be used on the shoe or the rail. Refractory coatings such as tantalum, nickel-aluminum, zirconium oxide, tungsten, and cobaltech have been used on rocket sleds in the past [11]. Other coating materials have been applied to the rails using sprayers. Although the application process is tedious and requires care, it has been somewhat successful in mitigating gouging. Currently, the HHSTT uses an iron oxide (hematite) coating on the low-speed section of the track and an epoxy coating on the latter half.

The second coating type is ablative, such as Teflon®, carbon-carbon, and carbon-phenolic coatings. These coatings have been used for speed exceeding Mach 6 with some success. Unfortunately, these materials do not offer good shock resistance (i.e., they fracture under shock loading), and they can cause detrimental configuration changes [11].

Both Barker et al. [22] and Tachau [25] proposed redesigned shoes based on varying the material properties of the shoes in order to reduce shock-wave effects and reduce the likelihood of creation of the high-pressure gouging core. These efforts were focused on system materials, whereas Schmitz's work [30] centered on gouge mitigation using coatings on the existing system materials.

Schmitz performed a study using CTH on various coatings and thicknesses and compared these conditions to the gouging threshold velocity. His results demonstrated that coatings made of aluminum, epoxy, polyethylene, polyurethane, and Teflon® raised the threshold gouging velocity substantially more than the other coatings including hematite, molybdenum, and zinc. These results from Schmitz et al.'s work appear in Figs. 2.11 and 2.12. Schmitz's definition of gouging, however, was limited to the creation and growth of the high-pressure core discussed earlier in this chapter.

A number of approaches have been attempted to mitigate gouging. The use of coatings and redesign of the sled's shoes are among these approaches. With

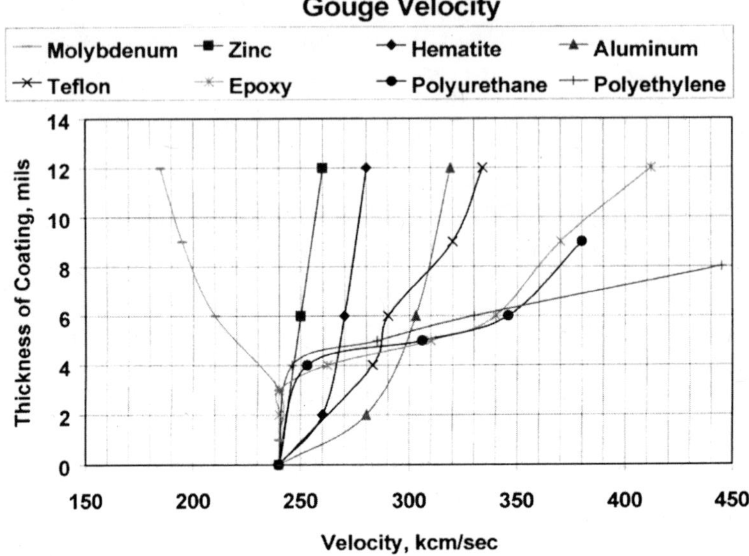

Fig. 2.11 CTH analysis of various coatings/thicknesses vs gouging velocity.

a clearer understanding of the mechanisms that initiate gouging and allow it to develop, more effective mitigation schemes can be developed. Certainly the past research has indicated that coating material choice and thickness, as well as impact geometry, play key roles in this field.

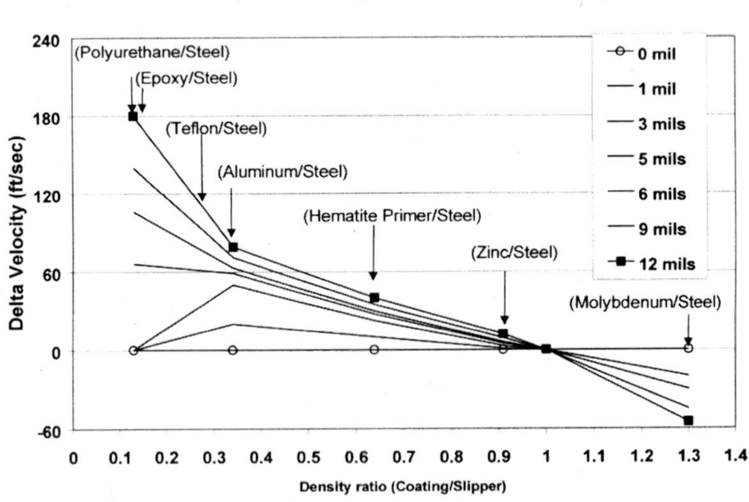

Fig. 2.12 CTH analysis of various coatings effectiveness.

IV. Szmerekovsky Model

Szmerekovsky and colleagues [2, 33–36, 49] advanced the understanding of the hypervelocity gouging problem by investigating the most appropriate numerical simulation tool for hypervelocity gouging research, performing a mesh refinement study and creating a gouging model in CTH, developing a mathematically sound technique to scale the shoe/sled model, examining some of the thermodynamics in the gouging problem, demonstrating by using a scaled approach that higher horizontal velocity created conditions are more conducive to gouging, and examining coating material selection and thickness, resulting in a proposal for application parameters to mitigate gouging.

In an effort to confront the gouging problem in general and the HHSTT geometry as a specific example, Szmerekovsky first evaluated a number of the available numerical codes to ascertain which one was the most appropriate for gouging study. Classic finite element formulations with Lagrangian meshes were found to lack the capability to deform and allow the material mixing seen in actual gouges. Additionally, the need to include thermodynamics and an extensive EOS database drove Szmerekovsky to choose CTH as the best code for this analysis. Although there are Eulerian codes that might have similar capabilities available, CTH provides unparalleled EOS data and shock physics capabilities that make it uniquely suited for this research.

Upon selecting CTH, the author constructed a model of the shoe and rail interaction problem at the HHSTT. As discussed in Sec. III, a plane strain (two-dimensional) model was considered a valid representation of the impact problem (see also Sec. II of Chapter 8). In addition, a three-dimensional model in CTH was not computationally feasible. Therefore a plane-strain model was created. Szmerekovksy then performed a mesh refinement study and discovered solution convergence at cell sizes of 0.002 cm. This also matched the material length scales described by Abu-Al-Rub and Voyiadjis [44] and Voyiadjis and Abu-Al-Rub [41], indicating that mesh sizes smaller than that would be outside the realm of continuum mechanics. The sled/rail model was created using the entire mass of the sled system, distributed to one shoe and depicted in plain strain. This included significantly more momentum than the Laird [1] model.

Szmerekovsky also applied the Buckingham pi theorem to the gouging problem to ensure a mathematically sound method of scaling the problem. He showed his model could replicate the results of Laird and that he could create a dimensionally accurate model of the shoe/rail interaction (in real dimensions) at the HHSTT.

Szmerekovsky's analysis included an initial look at the thermodynamics of the impact event. He concluded that friction and plasticity could generate the temperatures on the order of those needed to make the material phase changes and shear bands observed by Gerstle and colleagues [7–9]. Szmerekovsky also noted that CTH's heat conduction algorithm had the capability to model heat transport during the impact event away from the zones of heat generation into the bulk material. However, this effect was small and did not contribute significantly to the overall solution [49].

By applying the Buckingham pi scaling between impact scenarios at 1.5 and 3.0 km/s, he demonstrated that the 3.0 km/s impact generated gouging earlier than a scaled time history would predict. This showed that the characteristics leading to gouging were enhanced by the increased velocity. His model also created the typical high-pressure core necessary for gouge initiation and development.

Finally, Szmerekovsky concluded that coating selection and thickness could reduce shear stresses and pressure under the shoe and thereby reduce the probability of gouging. Additionally, the coating could act as a sacrificial layer that would effectively mask rail roughness (surface discontinuities or asperities) and allow the shoe to avoid a gouging scenario.

There are, however, several limitations to the work of Szmerekovsky and some areas that required further study. Primary among these is that CTH did not have the two specific materials, VascoMax 300 and 1080 steel, in its constitutive model material database. CTH has an EOS model for VascoMax 300 and an EOS model for iron (which can be argued to be very close to a low carbon steel like 1080 in terms of state relationships). However, neither material has a strength model in CTH. In addition, the previous models that were used relied on a strain-rate independent strength model, which poorly reflects the ability of the material to support higher stress levels while deforming at high rates. The fact that CTH used a strain-rate independent formulation was not known to Szmerekovsky. Second, the CTH model of the sled impact was never validated against experimental tests or shown to be accurate in its depiction of hypervelocity impact. Finally, available gouges from the HHSTT were not evaluated in terms of microstructure in order to quantify the thermal effects during gouging. A more thorough investigation of those thermal characteristics in the CTH environment would be helpful to both validate the model and to more fully understand the gouging process to devise mitigation methods.

V. Summary

The field of hypervelocity gouging has experienced a rich history of research. These efforts have ranged from experimentation on test tracks and laboratory examinations to numerical analysis, aerodynamic effects, load and failure analysis, and gouge mitigation. In these various approaches, a number of mechanisms for the initiation of gouging and the continued development of the gouge during the impact have been postulated.

Applying the conclusions of previous research to the HHSTT sled problem, one must consider the following areas: 1) experimental examination of hypervelocity gouges to ascertain characteristics of this kind of deformation; 2) close investigation of the thermodynamic effect on material microstructure to include the creation of shear bands caused by high thermal load and consideration of nonequilibrium thermodynamic effects; 3) material properties of the sled and rail, considering their ability to resist gouge initiation (specifically, the material flow models must be determined for the materials in the HHSTT impact scenario); 4) examination of the coatings used at the HHSTT to determine their ability to reduce friction in sliding impacts; 5) integration of as much of these phenomenon into a numerical simulation that can be validated using experimental data and utilized to provide additional insight into the mitigation of gouging; and 6) creation of a sled model that can be utilized to examine the various causes of gouge initiation (i.e., vertical impact, angled impact, and rail discontinuity).

These areas are ones in which further research must be done to characterize the gouging phenomenon and allow judgments to be made concerning gouge mitigation.

References

[1] Laird, D. J., "The Investigation of Hypervelocity Gouging AFIT/DS/ENY 02-01," Ph.D. Dissertation, Department of Aeronautics and Astronautics, Air Force Inst. of Technology, Wright-Patterson AFB, Dayton, OH, March 2002.

[2] Szmerekovsky, A. G., "The Physical Understanding of the Use of Coatings to Mitigate Hypervelocity Gouging Considering Real Test Sled Dimensions AFIT/DS/ENY 04-06," Ph.D. Dissertation, Department of Aeronautics and Astronautics, Air Force Inst. of Technology, Wright-Patterson AFB, Dayton, OH, Sept. 2004.

[3] Allvac, Inc., "VASCOMAX® C-200/C-250/C-300/C-350," Technical Data Sheet NI 253/154/156/173, Monroe, NC, 2000, www.allvac.com.

[4] MatWeb Material Property Data, "ANSI 1080 Steel," Technical Data Sheet, www.matweb.com, Sept. 2005.

[5] Zukas, J. A., *Introduction to Hydrocodes*, Studies in Applied Mechanics, No. 49, Elsevier, Oxford, U.K., 2004.

[6] Anderson, C. F., "An Overview of the Theory of Hydrocodes," *International Journal of Impact Engineering*, Vol. 5, Issues 1–4, 1987, pp. 33–59.

[7] Gerstle, F. P., "The Sandia Rocket Sled Rail and Slipper Study," Technical Rept., Sandia National Labs., Albuquerque, NM, 22 Aug. 1968.

[8] Gerstle, F. P., "Deformation of Dry Steel Surfaces During High-Velocity Sliding Contact," Ph.D. Dissertation, School of Engineering, Duke Univ., Durham, NC, June 1972.

[9] Gerstle, F. P., Follansbee, P. S., Pearsall, G. W., and Shepard, M. L., "Thermoplastic Shear and Fracture of Steel During High-Velocity Sliding," *Wear*, Vol. 24, Issue 1, 1973, pp. 97–106.

[10] Meyers, M. A., *Dynamic Behavior of Materials*, Wiley, New York, 1994.

[11] Krupovage, D. J., and Rasmussen, H. J., "Hypersonic Rocket Sled Development," High Speed Test Track Facility, 6585th Test Group, Technical Rept. JON: 99930000, Holloman AFB, NM, Sept. 1982.

[12] Krupovage, D. J., "Rail Gouging on the Holloman High Speed Test Track," 6585th Test Group, Test Track Div., Holloman AFB, NM, Jan. 1984.

[13] Barber, J. P., and Bauer, D. P., "Contact Phenomena at Hypervelocities," *Wear*, Vol. 78, Issues 1–2, 1982, pp. 163–169.

[14] Barber, J. R., and Ciavarella, M., "Contact Mechanics," *International Journal of Solids and Structures*, Vol. 37, 2000, pp. 29–43.

[15] Graff, K. F., and Dettloff, B. B., "The Gouging Phenomenon Between Metal Surfaces at Very High Speeds," *Wear*, Vol. 14, 1969, pp. 87–97.

[16] Mixon, L. C., "Assessment of Rocket Sled Slipper Wear/Gouging Phenomena," Applied Research Associates, Inc., Rept. F08635-97-C-0041, Albuquerque, NM, 1997.

[17] Graff, K. F., Dettloff, B. B., and Bolbulski, H. A., "Study of High Velocity Rail Damage," Air Force Special Weapons Center, Technical Rept. AA27 F29600-67-C-0043, Kirtland AFB, NM, Aug. 1970.

[18] Tarcza, K. R., and Weldon, W. F., "Metal Gouging at Low Relative Sliding Velocities," *Wear*, Vol. 209, Issues 1–2, 1997, pp. 21–30.

[19] Tarcza, K. R., "The Gouging Phenomenon at Low Relative Sliding Velocities," Master's Thesis, Department of Mechanical Engineering, Univ. of Texas at Austin, TX, Dec. 1995.

[20] Ramjaun, D., Kato, I., Takayama, K., and Jagadeesh, G., "Hypervelocity Impacts on Thin Metallic and Composite Space Debris Bumper Shields," *AIAA Journal*, Vol. 41, No. 8, Aug. 2003, pp. 1564–1572.

[21] Boehman, L. I., Barker, J. P., and Swift, H. F., "Simulation of Friction, Wear, Anti Gouging for Hypersonic Guider-Rail System," Defense Technical Information Rept., ADA054997, Fort Belvoir, VA, May 1978.

[22] Barker, L. M., Trucano, T. G., and Munford, L. W., "Surface Gouging by Hypervelocity Sliding Contact," Sandia National Labs., Technical Rept. SAND87-1328, Albuquerque, NM, Sept. 1987.

[23] Barker, L. M., Trucano, T. G., and Susoeff, A. R., "Railgun Rail Gouging by Hypervelocity Sliding Contact," *IEEE Transactions on Magnetics*, Vol. 25, No. 1, 1988, pp. 83–87.

[24] Susoeff, A. R., and Hawke, R. S., "Mechanical Bore Damage in Round Bore Composite Structure Railguns," Lawrence Livermore National Lab., Technical Rept. UCID-21520, Livermore, CA, 10 Feb. 1988.

[25] Tachau, R. D. M., "An Investigation of Gouge Initiation in High-Velocity Sliding Contact," Sandia National Labs., Technical Rept. SAND91-1732, Albuquerque, NM, Nov. 1991.

[26] Abrahamson, G. R., and Goodier, J. N., "The Hump Deformation Preceding a Moving Load on a Layer of Soft Material," *Journal of Applied Mechanics*, 61-APMW-5, Issue 5, 1961, pp. 12–26.

[27] Hooser, M., "Simulation of a 10,000 Foot Per Second Ground Vehicle," AIAA Paper 2000-2290, June 2000.

[28] Tachau, R. D. M., Yew, C. H., and Trucano, T. G., "Gouge Initiation in High-Velocity Rocket Sled Testing," *International Journal of Impact Engineering*, Vol. 17, Issues 4–6, 1995, pp. 825–836.

[29] Tachau, R. D. M., Yew, C. H., and Trucano, T. G., "Gouge Initiation in High-Velocity Rocket Sled Testing," Sandia National Labs., Technical Rept. SAND94-1333C, Albuquerque, NM, Oct. 1994.

[30] Schmitz, C. P., Palazotto, A. N., and Hooser, M., "Numerical Investigation of the Gouging Phenomena Within a Hypersonic Rail-Sled Assembly," AIAA Paper 2001-1191, April 2001.

[31] Laird, D. J., Palazotto, A. N., and Hooser, M. D., "High Speed Test Track Slipper/Rail Gouging Phenomena Simulations," *Thermal Hydraulics, Liquid Sloshing, Extreme Loads, and Structural Response*, 2001, pp. 61–68.

[32] Laird, D., and Palazotto, A., "Gouge Development During Hypervelocity Sliding Impact," *International Journal of Impact Engineering*, Vol. 30, Issue 2, Feb. 2004, pp. 205–223.

[33] Szmerekovsky, A. G., and Palazotto, A. N., "Structural Dynamics Considerations for a Hydrocode Analysis of Hypervelocity Test Sled Impacts," *AIAA Journal*, Vol. 44, No. 6, June 2006, pp. 1350–1359.

[34] Szmerekovsky, A. G., Palazotto, A. N., and Baker, W. P., "Scaling Numerical Models for Hypervelocity Test Sled Slipper-Rail Impacts," *International Journal of Impact Engineering*, Vol. 32, No. 6, 2006, pp. 928–946.

[35] Szmerekovsky, A. G., Palazotto, A. N., and Ernst, M. R., "Numerical Analysis for a Study of the Mitigation of Hypervelocity Gouging," AIAA Paper 2004-1922, April 2004.

[36] Szmerekovsky, A. G., Palazotto, A. N., and Ernst, M. R., "Numerical Analysis for a Study of the Mitigation of Hypervelocity Gouging," *Dayton-Cincinnati Aerospace Symposium*, modified and published as AIAA Paper 2004-1922, March 2004.

[37] Korkegi, R. H., and Briggs, R. A., "Aerodynamics of Hypersonic Slipper Bearing," Aerospace Research Lab., Technical Rept. ARL 68-0028, Wright-Patterson AFB, Dayton, OH, Feb. 1968.

[38] Korkegi, R. H., and Briggs, R. A., "The Hypersonic Slipper Bearing—A Test Track Problem," *Journal of Spacecraft*, Vol. 6, No. 2, 1969, pp. 210–212.

[39] Lofthouse, A. J., Hughson, M. C., and Palazotto, A. N., "Hypersonic Test Sled External Flow Field Investigation Using Computational Fluid Dynamics," AIAA Paper 2002-0306, Jan. 2002.

[40] Voyiadjis, G. Z., and Abed, F. H., "Microstructural Based Models for bcc and fcc Metals with Temperature and Strain Rate Dependency," *Mechanics of Materials*, Vol. 37, No. 2–3, 2005, pp. 355–378.

[41] Voyiadjis, G. Z., and Abu-Al-Rub, R. K., "Gradient Plasticity Theory with a Variable Length Scale Parameter," *International Journal of Solids and Structures*, Vol. 42, No. 14, 2005, pp. 3998–4029.

[42] Voyiadjis, G. Z., Abu-Al-Rub, R. K., and Palazotto, A. N., "Non-Local Coupling of Viscoplasticity and Anisotropic Viscodamage for Impact Problems Using the Gradient Theory," *Archives of Mechanics*, Vol. 55, No. 1, 2003, pp. 39–89.

[43] Voyiadjis, G. Z., Abu-Al-Rub, R. K., and Palazotto, A. N., "Thermodynamic Framework for Coupling of Non-Local Viscoplasticity and Non-Local Anisotropic Viscodamage for Dynamic Localization Problems Using Gradient Theory," *International Journal of Plasticity*, Vol. 20, No. 6, 2004, pp. 981–1038.

[44] Abu-Al-Rub, R. K., and Voyiadjis, G. Z., "Analytical and Experimental Determination of the Material Intrinsic Length Scale of Strain Gradient Plasticity Theory from Micro- and Nano-Indentation Experiments," *International Journal of Plasticity*, Vol. 20, No. 6, June 2004, pp. 1139–1182.

[45] Hanagud, S., "Thermomechanics of Impact and Penetration of Metallic Projectile into Isotropic and Granular Media," Presentation to Drs. Mook, Hughes, and Palazotto, Air Force Inst. of Technology, Wright-Patterson AFB, Dayton, OH, Oct. 2002.

[46] Jiang, T., Lu, X., and Hanagud, S. V., "Time-Dependent Penetration Model for High-Velocity Impact and Penetration: Phase Transition Studies," AIAA Paper 2005-2356, April 2005.

[47] Lu, X., and Hanagud, S. V., "Dislocation-Based Plasticity at High Strain-Rate in Solids," AIAA Paper 2004-1920, April 2004.

[48] Zukas, J. A., Nicholas, T., Swift, H. F., Greszczuk, L. B., and Curran, D. R., *Impact Dynamics*, Krieger, Malabar, FL, 1992.

[49] Szmerekovsky, A. G., Palazotto, A. N., and Cinnamon, J. D., "An Improved Study of Temperature Changes During Hypervelocity Sliding High Energy Impact," AIAA Paper 2006-2090, May 2006.

Chapter 3

Theoretical Background

I. Introduction

AS THE previous investigations indicated, the gouging phenomenon is dominated by a viscoplastic material deformation and shock-wave phenomenon. Therefore, attempts to accurately model these hypervelocity impacts revolve around determination of accurate material constitutive models and equations of state (EOS). Inherent in these two topics are the foundational elements upon which the solutions of these impacts rest. In addition, a brief discussion concerning the failure criterion within these solutions will be essential to understanding. Although it appears counterintuitive at first, one can begin the topic by outlining how a computer code, such as CTH, solves these types of problems. Of course, the solution techniques were derived prior to code implementation, but a discussion on the code solution procedure will establish a framework upon which we can rely to guide the discourse on theory.

II. Hypervelocity Impact Solution Procedure

Szmerekovsky in [1] outlines the theoretical basis for viscoplasticity and summarizes the key points that are germane to the development of a solution to the type of material deformations discussed in Chapter 2. He then describes how, in general, CTH solves such a problem. A summary of that description is provided in this section for ease of reference.

CTH, like most hydrocodes, uses the three conservation equations (mass, momentum, and energy), a description of the material EOS, and the constitutive relationships to solve the forcing function [2]. These three fundamental equations can be expressed in Lagrangian (material) or Eulerian (spatial) frames of reference [3]. A complete treatment of this can be found in Malvern's text [4], and a more specific application to these kind of problems can be found in [5] and [6].

CTH solves these high-energy impact problems by performing a Lagrangian step in which the material mesh is allowed to deform. The rationale for this is that the solution in the Lagrangian sense avoids the difficulty in solving the convective portion of the Eulerian representation. That is, the equations that must be solved to resolve material flow through the Eulerian (spatial) mesh are significantly more difficult than solving the material flow in the Lagrangian (material) sense. In terms of computational time, the difference between the approaches is significant [3]. After the deformation is computed in the material sense, the deformed material is then re-mapped back to an Eulerian mesh, and the time step is complete. Using this technique, the material distribution is first solved using the Lagrangian coordinate system and then mapped back into the Eulerian coordinate system [3]. The relationships required for this process are presented here.

In the following equations, the summation convention is adopted for the repeated indices. The Lagrangian (material) expression of the conservation of mass in terms of measurements in the Eulerian (spatial) coordinate system is

$$\frac{\mathrm{D}\rho}{\mathrm{D}t} + \rho\frac{\partial v_i}{\partial x_i} = 0 \tag{3.1}$$

where ρ the material density, v_i the velocity (evaluated in the spatial or Eulerian coordinate system for a specific particle), x a measure of position (in the spatial coordinate system), t time, and

$$\frac{\mathrm{D}}{\mathrm{D}t} = \frac{\partial}{\partial t} + v_i\frac{\partial}{\partial x_i} \tag{3.2}$$

which is known as the material derivative, substantial derivative, or the total time derivative (in which the measurements of position and velocity are in the Eulerian or spatial coordinate system). This definition establishes how a Lagrangian (material) description can be formed from quantities measured in the Eulerian (spatial) coordinate system.

The conservation of mass can therefore be expressed in the Eulerian reference frame, in which we are tracing density as it flows through a specific point, as

$$\frac{\partial\rho}{\partial t} + \frac{\partial}{\partial x_i}(\rho v_i) = 0 \tag{3.3}$$

The conservation of momentum can be expressed also, with σ_{ij} representing the stress tensor and f_i denoting the external body forces per unit mass. In the Lagrangian (material) sense, the relation is

$$\frac{\mathrm{D}v_i}{\mathrm{D}t} = f_i + \frac{1}{\rho}\frac{\partial\sigma_{ji}}{\partial x_j} \tag{3.4}$$

and in the Eulerian frame it becomes

$$\frac{\partial v_i}{\partial t} + v_j\frac{\partial v_i}{\partial x_j} = f_i + \frac{1}{\rho}\frac{\partial\sigma_{ji}}{\partial x_j} \tag{3.5}$$

The conservation of energy, with e representing the specific total energy, can be expressed in the Lagrangian (material) sense as

$$\frac{\mathrm{D}e}{\mathrm{D}t} = f_i v_i + \frac{1}{\rho}\frac{\partial}{\partial x_j}(\sigma_{ij}v_i) \tag{3.6}$$

and in the Eulerian frame it can be written as

$$\frac{\partial e}{\partial t} + v_i\frac{\partial e}{\partial x_i} = f_i v_i + \frac{1}{\rho}\frac{\partial}{\partial x_j}(\sigma_{ij}v_i) \tag{3.7}$$

The total specific energy is defined as

$$e = \frac{1}{2} v_i v_i + E \tag{3.8}$$

which comprises the kinetic energy and the specific internal energy E. With this expression, Eq. (3.6) becomes

$$\frac{\mathrm{D}E}{\mathrm{D}t} = \frac{P}{\rho^2} \frac{\mathrm{D}\rho}{\mathrm{D}t} + \frac{1}{\rho} s_{ij} \dot{\varepsilon}_{ij} \tag{3.9}$$

and Eq. (3.7) becomes

$$\frac{\partial E}{\partial t} + v_i \frac{\partial E}{\partial x_i} = \frac{P}{\rho^2} \left(\frac{\partial \rho}{\partial t} + v_i \frac{\partial \rho}{\partial x_i} \right) + \frac{1}{\rho} s_{ij} \dot{\varepsilon}_{ij} \tag{3.10}$$

where s_{ij} is the stress deviator tensor, $\dot{\varepsilon}_{ij}$ is the strain-rate tensor, and P is the hydrostatic pressure.

Another way to state this is that the total stress tensor is considered in its two components, the symmetric deviatoric stress tensor $\boldsymbol{D}$, and the spherical stress tensor, $\boldsymbol{S}$ (in which the stress tensor is spatial or Cauchy stress).

These conservation equations are not sufficient to solve the deformation problem. Two additional equations are necessary. The first is the equation of state, which relates the hydrostatic pressure to state variables such as density and specific internal energy. The other is the material constitutive model, which describes the material flow stress as a function of strain, strain rate, and temperature.

The standard solution method is to allow the EOS to provide the solution for the spherical (sometimes referred to as volumetric) stress and the constitutive model to provide the solution to the deviatoric (sometime referred to as shear) stress.

In hypervelocity impact, where impacts occur at speeds on the order of magnitude of the material sound speed, the spherical stresses tend to be much higher than the deviatoric stresses. Therefore, it is easy to err and assume that the EOS is the primary consideration for creating accurate solutions. However, even in these high energy impacts, the pressures quickly drop (especially away from the impact interface) to regimes in which the deviatoric stresses dominate. Therefore, an accurate model must have both a robust EOS and an accurate constitutive model to generate good results [2].

III. Equation of State

Stated simply, the equation of state bridges traditional continuum mechanics and thermodynamics. Continuum mechanics solves the three conservation equations (in terms of pressure P, density ρ, energy E, and particle velocity u) in a continuous field with respect to time t and space. Of course, these three relationships are incomplete—requiring a fourth relation. Continuity equations can provide the missing relationships in traditional static mechanics. In this high-energy regime, an equation of state $P(\rho, E)$ can also provide the required fourth relationship [3, 7–9].

Let us assume, for the moment, that $P(\rho, E)$ is a unique function. This implies that ρ and E are state variables and P is a state function. If this is true, then P is independent of the process that generated the conditions and is only a function of the $\rho - E$ state.

If these state variables exists and are unique, then the process of thermodynamics can be applied to the problem. If we consider an equilibrium thermodynamic condition, the other thermodynamic quantities must also exist (i.e., temperature T and entropy S). The full complement of state equations would then become $P(\rho, T)$, $E(\rho, T)$, and $S(\rho, T)$. This set is sometimes referred to as the temperature-based EOS, as opposed to the energy-based formulation just presented.

Pressure and internal energy can be derived from the Helmholtz free energy relationship, which states

$$A = E - TS \tag{3.11}$$

where A is the Helmholtz free energy. The pressure and internal energy can then be expressed by

$$P = -\left(\frac{\partial A}{\partial V}\right)_T = \rho^2\left(\frac{\partial A}{\partial \rho}\right)_T \tag{3.12}$$

$$E = A - T\left(\frac{\partial A}{\partial T}\right)_\rho \tag{3.13}$$

where $V = 1/\rho$. Combining Eqs. (3.12) and (3.13), we arrive at the following relationship (sometimes referred to as the thermodynamic consistency relation):

$$-\left(\frac{\partial E}{\partial V}\right)_T = \rho^2\left(\frac{\partial E}{\partial \rho}\right)_T = P - T\left(\frac{\partial P}{\partial T}\right)_\rho \tag{3.14}$$

Once again, this formulation assumes a state of thermodynamic equilibrium. Note that in Eq. (3.12), the work done by $P\,\mathrm{d}V$ does not exclusively go into strain energy. Some of this work is converted into entropy, or heat. If we were considering an isentropic problem, then the equation would reduce (noting $\mathrm{d}S = 0$ and $-P\,\mathrm{d}V = \mathrm{d}E$) to

$$P = -\left(\frac{\partial E}{\partial V}\right)_S = \rho^2\left(\frac{\partial E}{\partial \rho}\right)_S \tag{3.15}$$

Now, let us consider a situation in which $P(\rho, E)$ is not a unique solution. Experimentation provides us evidence that this is true. Therefore, to create an EOS, more variables must be introduced to create a unique solution. Two primary cases apply to our problem. (Others involving explosive products are beyond this investigation.) One is that the material's microstructure can have an impact on the EOS (like grain boundaries, defects, phase changes, etc.), requiring the addition of flow variables that average the effect of inhomogeneous microstructure. The second case is that of time-dependent behavior in which the time required to reach equilibrium is not available in the given problem. This tends to "overdrive" the EOS into a state of thermodynamic nonequilibrium until the required time to reach equilibrium has been satisfied. Rapid deformation and heating can cause this second case.

To extend the concept of the EOS to handle these additional cases (which apply to the HHSTT problem, described in detail later), some additional variables need to be considered. These variables are sometimes referred to as "internal state variables." This is because they do not appear in the conservation equations. These internal state variables modify the EOS relationships—specifically the thermodynamic relations—to account for these other effects.

Complicating this matter further are impacts in which shock waves are generated. Szmerekovsky [1] details the basics of shock waves. Briefly, a shock wave is generated when an impact creates disturbances that propagate faster than the material speed of sound. A sharp discontinuity is created in which material states vary significantly across a moving shock wave. In this case, the solution must consider the Rankine–Hugoniot jump conditions [9], where the states across the wave front must relate via the shock conservation equations (here the subscript 0 refers to the initial, at rest condition):

$$\mu = 1 - \frac{\rho_0}{\rho} = \frac{u_P}{U_S} \tag{3.16}$$

$$P_H = P_0 + \rho_0 U_S u_P = P_0 + \rho_0 U_S^2 \mu \tag{3.17}$$

$$E_H = E_0 + \frac{1}{2}u_P^2 = E_0 + \frac{(P_H + P_0)\mu}{2\rho_0} \tag{3.18}$$

where ρ is the density, P_H is the pressure, E_H is the entropy of the shocked material, U_S is the shock wave velocity through the undisturbed material, and u_P is the particle velocity behind the shock front. These jump conditions must also be satisfied by the EOS in solutions consider shock waves.

Beyond these considerations are those that advanced EOS' handle. To this point, a state of nonequilibrium thermodynamics could be achieved by creating rapid deformation and heating [3, 10, 11]. In fact, many mechanisms exist that can create this state of nonequilibrium. Some of them that apply to hypervelocity impact are material phase changes that nonconservatively remove energy from the solution and generate multiple shock fronts, thermal electronic excitation and ionization, and nonconstant heat capacity. Other mechanisms, such as chemical reactions, are not contributors to the HHSTT problem. Recognizing that these advanced considerations must be accounted for in the EOS (typically through internal state variables), the EOS becomes much more involved [11].

Because the creation of a single EOS formulation that encompasses a large body of materials remains elusive, equations of state tend to be developed for a specific material, and even specific impact conditions. These EOS formulations are then validated against experimentation (high-energy impact).

Fortunately, CTH not only possesses many analytical EOS formulations, with the appropriate constants for specific materials, but it also includes a very powerful, and frankly unique, database of experiments for many materials. These experiments have measured the state variables for hypervelocity impacts, explosions, and other high-energy events. The states are recorded in a table that CTH can reference in the solution procedure. These tables create a solution surface in state variable space that includes all of the higher-order effects just noted. In essence, the EOS is not so much an equation as a state look-up table. Therefore, all of the advanced conditions and cases are inherently included because of the nature of the experimentation.

For the materials in the HHSTT gouging problem, CTH has EOS tables for VascoMax 300 and iron. The iron EOS is considered very accurate for low carbon steels, like 1080 steel, and includes the phase transitions that occur at high pressure and temperature. Therefore, from an EOS point of view CTH is uniquely suited to solve the gouging problem. Additionally, the EOS tables include nonequilibrium thermodynamic effects that occur in these hypervelocity impacts.

With the EOS solving the volumetric (spherical) stress components of the impact, we require a constitutive model of the materials to handle the deviatoric portion of the deformation.

IV. Constitutive Models

To solve the deviatoric stress portion of the high-energy impact problem, a constitutive model is required. A constitutive model is a relationship defining the dynamic yield strength (also known as material flow stress) of the material. This is, of course, much different than the static yield strength of the same material. As the concept of the material constitutive model has developed over the last several decades, more aspects of the problem have been considered in the model. Most of the modern formulations of material flow models have recognized the effects of strain ε_{ij}, strain rate $\dot{\varepsilon}_{ij}$, and temperature [2]. This can be expressed as

$$\sigma_{ij} = f(\varepsilon_{ij}, \dot{\varepsilon}_{ij}, T) \tag{3.19}$$

Figure 3.1 illustrates the basic relationships. Increasing strain and strain rate lead to higher flow strength (strain hardening and strain-rate dependency respectively), whereas increasing temperature results in lower dynamic yield strength (thermal softening). Note that we are now discussing quantities known as "effective stress" and "effective strain/strain rate." This is because creating constitutive models that are functions of the entire stress tensor is very cumbersome. Therefore, an isotropic material is assumed, and stress is taken to be a function of scalar quantities [2].

There are numerous constitutive models available in the literature and within the CTH code. The major differences between them revolve around being able to manipulate the strength curve with respect to the just-mentioned parameters to create desired results, or on how linked the specific formulation is with material properties/characteristics. Some of them attempt to account for noncontinuum effects, such as material microstructure and defects. Others are simply empirically based flow models based on experimentation.

Two of the most widely used constitutive models are the Johnson–Cook and the Zerilli–Armstrong formulations. The strength of these particular models is the strong success that they have historically enjoyed in modeling material behavior, while being straightforward enough to be determined for a specific material without inordinate expense.

Previous modeling efforts by Szmerekovsky were based on the Steinburg–Guinan–Lund model [12–14] for VascoMax 250 and the Johnson–Cook model for iron. Unfortunately, CTH's implementation of the Steinburg–Guinan–Lund

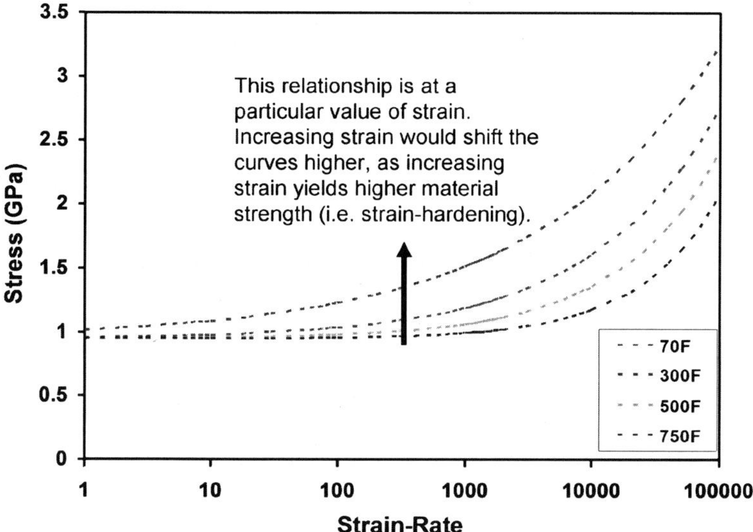

Fig. 3.1 General constitutive model relations.

model for VascoMax 250 was a rate-independent version, a fact not known to the researcher. Therefore, the results generated by his models are limited to a single value of flow stress (given constant temperature and strain) over the wide range of strain rates. Additionally, the Johnson–Cook model for iron is limited in its applicability to a relatively high-strength steel, such as 1080 steel, which has very different strain-rate dependency. These factors prompt this author to seek specific, verifiable material flow models for VascoMax 300 and 1080 steel.

The Johnson–Cook model was formulated first in 1983 and was proposed as predominately an empirical model [15]. Later work on the model highlights the fact that some of the constants show trends based on material properties and families of materials (i.e., all steels might share a certain range for a particular constant). The Johnson–Cook model relates the material flow stress (dynamic yield strength) σ as

$$\sigma = (A + B\varepsilon^n)(1 + C \, \ell n \, \dot{\varepsilon}^*)(1 - T^{*m}) \tag{3.20}$$

where ε is the equivalent plastic strain, $\dot{\varepsilon}^*$ is the dimensionless strain rate ($\dot{\varepsilon}^* = \dot{\varepsilon}/\dot{\varepsilon}_0$ and $\dot{\varepsilon}_0 = 1.0 \text{ s}^{-1}$), and T^* is the homologous temperature defined as

$$T^* = (T - T_{\text{room}})/(T_{\text{melt}} - T_{\text{room}}) \tag{3.21}$$

The constants A, B, C, m, and n are determined via experimental testing. The Johnson–Cook model is somewhat easier to experimentally determine because of the discrete parts that account for the various parameters. The first portion is the static yield strength and a modification for strain. The second portion adds the strain-rate dependency. The final portion adds the temperature effects. By performing experiments in a state of uniaxial strain, over a range of temperatures and strain rates, one can create the model for chosen values of strain.

The primary limitation to the Johnson–Cook model is that is creates a flow stress curve that is linear with respect to strain rate. That is, the curves represented in Fig. 3.1 cannot be produced by the Johnson–Cook model. Rather, those curves would be linear. For some applications, this model is sufficient—especially if the strain rates under consideration remain at 10^4/s or less.

The type of constitutive model that generated the curves in Fig. 3.1 is the Zerilli–Armstrong model. The Zerilli–Armstrong formulation was first proposed in 1987 [16]. This model is based more upon microstructural characteristics of the materials, yet still retains an empirical basis similar to the Johnson–Cook model. In the Zerilli–Armstrong model, the flow stress σ for a face-centered cubic (FCC) crystal lattice is

$$\sigma = \Delta\sigma'_G + c_2\sqrt{\varepsilon}\, e^{(-c_3T + c_4T\ \ell n\ \dot{\varepsilon})} + k\sqrt{\ell} \tag{3.22}$$

where $\Delta\sigma'_G$ is a stress component accounting for dislocation density, $k\sqrt{\ell}$ is an incremental stress term that includes the microstructural stress density k, and the average grain diameter ℓ. As with the Johnson–Cook formulation, ε is the effective plastic strain, $\dot{\varepsilon}$ is the effective plastic strain rate, and T is temperature. The terms c_2 through c_5 are experimentally determined for each specific material. The units of these constants must be reported, and the units of c_3 and c_4 must cancel the selected units of T. This formulation is modified slightly for the body-centered cubic (BCC) lattice material:

$$\sigma = \Delta\sigma'_G + c_1 e^{(-c_3T + c_4T\ \ell n\ \dot{\varepsilon})} + c_5\varepsilon^n + k\sqrt{\ell} \tag{3.23}$$

Note that the some of the terms were slightly modified. It is possible to combine Eqs. (3.22) and (3.23) into one expression, in which either $c_1 = 0$ for FCC materials or $c_2 = 0$ for BCC materials (which is the case for the two that are considered in this work). Additionally, one can combine the material specific terms that do not rely on experimentation into a single expression by allowing $A = \Delta\sigma'_G + k\sqrt{\ell}$. The combined expression then becomes

$$\sigma = A + (c_1 + c_2\sqrt{\varepsilon})e^{(-c_3T + c_4T\ \ell n\ \dot{\varepsilon})} + c_5\varepsilon^n \tag{3.24}$$

This is the formulation that will be used in this study. Therefore, the terms A, n, and c_1 through c_5 (with $c_2 = 0$) are those that will be determined for the specific materials.

These two constitutive models establish the viscoplastic flow stress for the materials under dynamic deformation. They both are strain-rate dependent and include the characteristics of strain hardening and thermal softening. These two particular constitutive models are the most widely used within the field because some straightforward experimentation can be employed to determine the model constants. Additionally, they have a history of generating accurate results if they are judiciously applied to impact problems.

With the deviatoric stresses being solved via the material constitutive relationships, the final element of the theoretical approach is the delineation of the failure model.

V. Failure Model

The topic of failure models and their application in the solution of hypervelocity impact, especially within computational codes, is a particularly wide-ranging topic [2, 9, 17]. There are many different approaches to the concept of defining the criteria (such as pressure, temperature, shear stress, etc.) that lead to material failure. Unfortunately, even the most complex theories, based on microstructural factors, are inadequate to apply across a range of materials [2]. There are very few that gain acceptance by the community of hypervelocity impact researchers and therefore become placed into the computational codes.

One of the most accurate approaches for the failure of metals, specifically, is the establishment of a hydrostatic failure pressure at which the material fractures. In the absence of accepted theory on the determination of this pressure, it is typically determined from experimentation. In this case, isentropic material properties are assumed, and an uniaxial fracture stress is used to estimate the failure pressure. The Von Mises failure criterion is the most common, and most accurate, for metals undergoing deformation [18]. Using the notation of this chapter,

$$(\sigma_1 - \sigma_2)^2 + (\sigma_2 - \sigma_3)^2 + (\sigma_3 - \sigma_1)^2 = 2\sigma^2 \tag{3.25}$$

where σ is the uniaxial flow stress and σ_1, σ_2, and σ_3 are the principal stresses.

CTH, in its solution procedure, uses this hydrostatic failure pressure, called the fracture pressure within the code, to delineate the failure criterion. Cells with mixed material will take on a volume-averaged failure pressure, unless specifically specified otherwise.

As discussed in Chapter 2, many contemporary researchers are attempting to create more advanced failure criteria and damage algorithms [19–25]. However, these approaches have not yet achieved widespread acceptance nor have they been made available within the major hydrocodes. Therefore, their utility is limited. Additionally, these investigators have not yet proven their theories, when applied to hypervelocity impact cases, will replicate experimental results with greater accuracy than the current approach.

Therefore, for the purpose of this study, the hydrostatic failure pressure approach will be utilized. The determination of the specific quantity of this failure pressure from experiment will be described in Chapter 5.

This brief theoretical overview, then, establishes the framework upon which this study rests. To accurately model hypervelocity gouging, we need to determine the constitutive models for VascoMax 300 and 1080 steel. The EOS for these materials within CTH is extremely accurate, experimentally based, and includes nonequilibrium thermodynamic characteristics. However, there are no accurate constitutive expressions for the material in question. Once we arrive at the constitutive models, a validation of CTH's ability to model impact scenarios needs to be conducted. Using this approach, we can be assured that we have created the most accurate computational model of the HHSTT problem.

Additional theory will be presented as it specifically applies to the study being discussed. In this way, the clarity of the approach will be maintained.

Before the constitutive models for VascoMax 300 and 1080 steel are determined, the characteristics of gouging needs to be explored via an experimental examination of a gouged rail from the HHSTT.

References

[1] Szmerekovsky, A. G., "The Physical Understanding of the Use of Coatings to Mitigate Hypervelocity Gouging Considering Real Test Sled Dimensions AFIT/DS/ENY 04-06," Ph.D. Dissertation, Department of Aeronautics and Astronautics, Air Force Inst. of Technology, Wright-Patterson AFB, Dayton, OH, Sept. 2004.

[2] Zukas, J. A., Nicholas, T., Swift, H. F., Greszczuk, L. B., and Curran, D. R., *Impact Dynamics*, Krieger, Malabar, FL, 1992.

[3] Anderson, C. F., "An Overview of the Theory of Hydrocodes," *International Journal of Impact Engineering*, Vol. 5, 22 Aug. 1987, pp. 33–59.

[4] Malvern, L. E., *Introduction to the Mechanics of a Continuous Medium*, Prentice–Hall, Upper Saddle River, NJ, 1969.

[5] Benson, D. J., "Computational Methods in Lagrangian and Eulerian Hydrocodes," *Computer Methods in Applied Mechanics and Engineering*, Vol. 99, Issues 2–3, Sept. 1992, pp. 235–394.

[6] Karpp, R. R., "Warhead Simulation Techniques: Hydrocodes," *Tactical Missile Warheads*, Progress in Astronautics and Aeronautics, Vol. 155, Seebass, A. R. (ed), AIAA, Washington, DC, 1993, pp. 223–313.

[7] Hertel, E. S., Bell, R. L., Elrick, M. G., Farnsworth, A. V., Kerley, G. I., McGlaun, J. M., Petney, S. V., Silling, S. A., Taylor, P. A., and Yarrington, L., "CTH: A Software Family for Multidimensional Shock Physics Analysis," *Proceedings of the 19th International Symposium on Shock Waves*, Sandia National Laboratories Rept. SAND-92-2089C, pp. 377–382.

[8] McGlaun, J. M., Thompson, S. L., and Elrick, M. G., "CTH: A Three-Dimensional Shock Wave Physics Code," *International Journal of Impact Engineering*, Vol. 10, Issues 1–4, 1990, pp. 351–360.

[9] Zukas, J. A., *Introduction to Hydrocodes*, Studies in Applied Mechanics, No. 49, Elsevier, Oxford, U.K., 2004.

[10] Anderson, J. D., Jr., *Modern Compressible Flow: with Historical Perspective*, 2nd ed., McGraw–Hill, New York, 1990.

[11] Anderson, J. D., Jr., *Hypersonic and High Temperature Gas Dynamics*, AIAA, Reston, VA, 2000.

[12] Steinberg, D., and Lund, C., "A Constitutive Model for Strain Rates from 10^{-4} to 10^{6} sec^{-1}," *Journal of Applied Physics*, Vol. 65, 1989, p. 1528.

[13] Steinberg, D. J., "Equation of State and Strength Properties of Selected Materials," Lawrence Livermore National Lab., Technical Rept. UCRL-MA-106439, Livermore, CA, Feb. 1996.

[14] Steinberg, D. J., Cochran, S. G., and Guinan, M. W., "A Constitutive Model for Metals Applicable at High-Strain Rate," *Journal of Applied Physics*, Vol. 51, No. 3, 1980, pp. 1498–1504.

[15] Johnson, G. R., and Cook, W. H., "A Constitutive Model and Data for Metals Subjected to Large Strains, High Strain Rates, and High Temperatures," *Proceedings of the 7th International Symposium Ballistics*, American Defense Preparation Organization, The Hague, The Netherlands, April 1983, pp. 541–547.

[16] Zerilli, F. J., and Armstrong, R. W., "Dislocation-Mechanics-Based Constitutive Relations for Material Dynamics Calculations," *Journal of Applied Physics*, Vol. 61, No. 5, 1987, pp. 1816–1825.

[17] Zukas, J. A., "Survey of Computer Codes for Impact Simulation," *High Velocity Impact Dynamics*, Wiley, New York, 1990, pp. 593–623.

[18] Broek, D., *Elementary Engineering Fracture Mechanics*, 4th ed., Kluwer Academic, Dordrecht, The Netherlands, 1986.

[19] Abu-Al-Rub, R. K., and Voyiadjis, G. Z., "Analytical and Experimental Determination of the Material Intrinsic Length Scale of Strain Gradient Plasticity Theory from Micro- and Nano-Indentation Experiments," *International Journal of Plasticity*, Vol. 20, Issue 6, June 2004, pp. 1139–1182.

[20] Hanagud, S., "Thermomechanics of Impact and Penetration of Metallic Projectile into Isotropic and Granular Media," Presentation to Drs. Mook, Hughes, and Palazotto, Air Force Inst. of Technology, Wright-Patterson AFB, Dayton, OH, Oct. 2002.

[21] Voyiadjis, G. Z., and Abed, F. H., "Microstructural Based Models for bcc and fcc Metals with Temperature and Strain Rate Dependency," *Mechanics of Materials*, Vol. 37, Nos. 2–3, 2005, pp. 355–378.

[22] Voyiadjis, G. Z., and Abu-Al-Rub, R. K., "Gradient Plasticity Theory with a Variable Length Scale Parameter," *International Journal of Solids and Structures*, Vol. 42, No. 14, 2005, pp. 3998–4029.

[23] Voyiadjis, G. Z., Abu-Al-Rub, R. K., and Palazotto, A. N., "Non-Local Coupling of Viscoplasticity and Anisotropic Viscodamage for Impact Problems Using the Gradient Theory," *Archives of Mechanics*, Vol. 55, No. 1, 2003, pp. 39–89.

[24] Voyiadjis, G. Z., Abu-Al-Rub, R. K., and Palazotto, A. N., "Thermodynamic Framework for Coupling of Non-Local Viscoplasticity and Non-Local Antisotropic Viscodamage for Dynamic Localization Problems Using Gradient Theory," *International Journal of Plasticity*, Vol. 20, No. 6, 2004, pp. 981–1038.

[25] Voyiadjis, G. Z., and Abu-Al-Rub, R. K., "Non-Local Coupling of Viscoplasticity and Rate-Dependent Damage for Impact Problems," Dept. of Civil and Environmental Engineering, Technical Rept. 1, Louisiana State Univ., Baton Rouge, LA, Aug. 2002.

Chapter 4

Characterization of Gouging

I. Introduction

TO MORE fully understand the phenomenon of hypervelocity gouging, a section of gouged rail from the HHSTT was examined. As described in Chapter 2, the sled experiences catastrophic deceleration at the end of a successful test. Therefore, gouged shoes are not available for examination.

The gouged rail was examined using the most modern techniques in order to understand the characteristics of gouging and to develop experimental elements for CTH to replicate as a part of the model validation process. The examination of the gouged rail revealed a thermal profile that can used to compare against simulation results [1, 2].

Based on the results of examining the gouged rail and the resulting evidence of thermally induced microstructure change, an examination of additional rail sections was undertaken. A slightly damaged section and three sections of undamaged rail were investigated to determine if microstructure changes are evident in these as well.

II. Methodology for Examining Gouge

The HHSTT has changed materials for the hypervelocity sled's shoes many times in an effort to mitigate gouging. For example, in 1973 [3], the rail (fabricated of ANSI 1080 steel) and the shoes (at that time constructed of 304 and 17-4 PH stainless steel) were examined following a 1.77-km/s test run. The microstructure of the damaged area was examined, and the authors concluded that shear bands and distorted pearlite comprised a majority of the subsurface effects. This treatment, however, was limited to optical microscopy and did not make an examination of the microstructure into the depth of the rail (away from the impact interface).

Similar to this previous work, a section of rail (featured in Fig. 1.5) damaged at approximately 2.1 km/s was sliced normal to the direction of motion to prepare specimens for metallographic analysis. The slice was removed from the center of the gouge, as measured in terms of the gouge's orientation down the rail, in the direction of the sled's motion. Figure 4.1 is of the slice of damaged rail, with the velocity vector of the sled shown. This rail section has an iron-oxide coating applied to aid in the mitigation of the gouging event.

The top of this rail section was removed and made into four separate specimens, denoted as specimens B1 through B4. An undamaged rail section was likewise

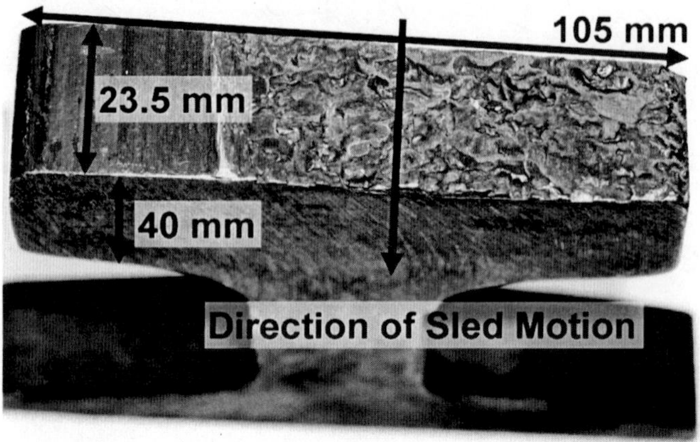

Fig. 4.1 Section of gouged rail.

sectioned to serve as the experimental control. Figure 4.2 indicates the orientation of these samples. For samples B1, B2, and B3, the face examined is the one pictured. The face examined for specimen B4 was the face shared with B3 or the plane of the specimen along the velocity vector of the sled (lengthwise down the gouge).

The samples were mounted in conducting material and polished to a 0.05-μ finish using standard techniques. The fine polishing was limited to using diamond. It was noted that if the samples were further polished using 0.05-μ colloidal silica, the samples partially etched, even though the preparation was at a neutral pH. The best results for examining the microstructure of the 1080 steel were obtained using an electropolishing technique instead of the silica as the last step. In addition, etched specimens (with a standard 3% nital etch) were prepared.

Therefore, three different preparations were available for examination. The first was created by polishing the specimens to a 0.5-μ finish with diamond and finishing them with silica, which will be referred to as the "as-polished" state.

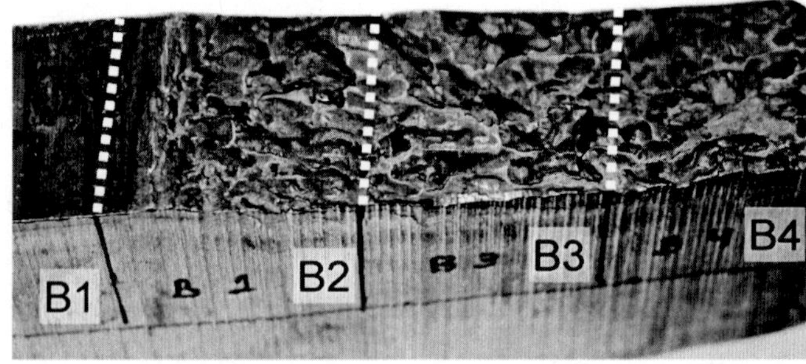

Fig. 4.2 Gouged rail specimen location (specimens are approximately cubic, with 25-mm sides).

Table 4.1 **Summary of specimen preparation/examination techniques**

Examination technique	Polished with diamond to 0.05 μ, finished with silica, partial etch	Polished with diamond to 0.05 μ, finished with silica, etched with 3% nital	Polished with diamond to 0.05 μ, electropolished
Optical microscopy, polarized light	X	—	X
Optical microscopy, bright field light	—	X	—
SEM, both primary and backscatter modes	X	—	X
EDS	X	—	—
OIM	—	—	X
Nomenclature	As-polished	Etched	Electropolished

The second was created by finishing them with an electropolish instead of silica, which will be referred to as "electropolished." Finally, the third preparation was created by etching the specimens finished with silica, which will be referred to as "etched."

The specimens were examined in polarized light, in bright field, and using a scanning electron microscope (SEM) as appropriate to the final finish. Micro hardness testing and energy dispersive x-ray spectroscopy (EDS) were performed on the specimen in the as-polished state and orientation imaging spectroscopy (OIM) in the electropolished state. Table 4.1 summarizes the various preparations and what examination techniques were applied and provides their nomenclature, which is used for the balance of this work. The facilities at the Air Force Material Laboratory, Materials Division, at Wright-Patterson Air Force Base were generously made available for this analysis.

III. Results of the Examination of Gouged Specimens

The examination of the gouge specimens indicated a large thermal event that permanently changed the microstructure of the steel. This change was evident to the naked eye, as well as through the many examination techniques applied.

Specimens in the as-polished state exhibited a partial etch of the surface that was visible to the naked eye. Putting a camera at a slight angle to the surface of the specimen and taking a flash photograph revealed a clear impact affected area. Not only is a plastically deformed region present, but also a zone that appears to be affected by the heat of the event (see Fig. 4.3) is present. Figure 4.4 provides a comparison of specimen B4 to specimen A4 (from the undeformed rail), and the differences are clear. The lower right section of B4 shows a crack that was present

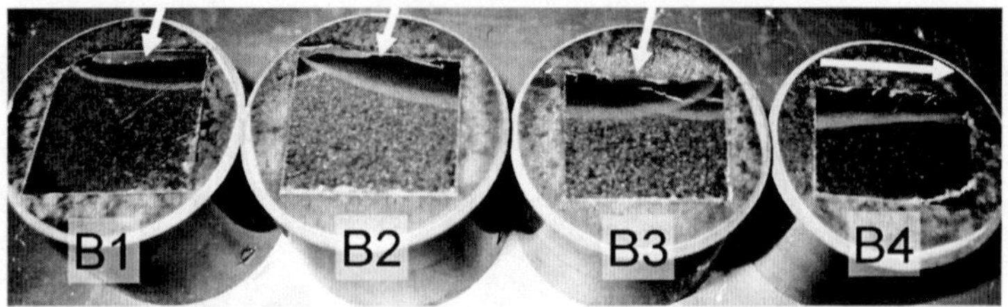

Fig. 4.3 Overview of heat-affected zone (velocity vector of sled shown by the arrows).

down to a rail depth of 31 mm. In addition, the removal of approximately 5 mm of material from the rail is evident in this figure.

Closer to the surface in the same specimen are some clear shear bands (further justified and described later in this work) and indications that the material flowed in the direction of the sled's travel to modify the top of the typical cracking in the gouged area that was reported in [3]. That is, the curvature of these shear bands being concave against the sled's motion in [3] is modified at the top to follow the plastic flow of the material. In specimen B4, seven such discrete bands are present (with similar shape) indicative of the material flowing close to the surface and thereby modifying the shear band curvature. This material flow indicates that the surface melted during gouging impact [3].

It is helpful, at this point, to note that all of the 1080 steel specimens contained inclusions and flaws in the microstructure as a result of the manufacturing. In addition, after metallurgical examination of all of the deformed rail specimens, it was concluded that all of their microstructures were significantly similar; therefore, the results for B4 are representative. Finally, where available, the microstructure reported was matched against published micrographs elsewhere in the literature (in particular in [4] and [5]).

When the entire specimen B4 is examined, the zone affected by the thermal characteristics of the hypervelocity impact is evident. Figure 4.5 shows B4 in an etched state (and is in the plane of motion examined by the computer modeling; see Chapter 2, Sec. IV).

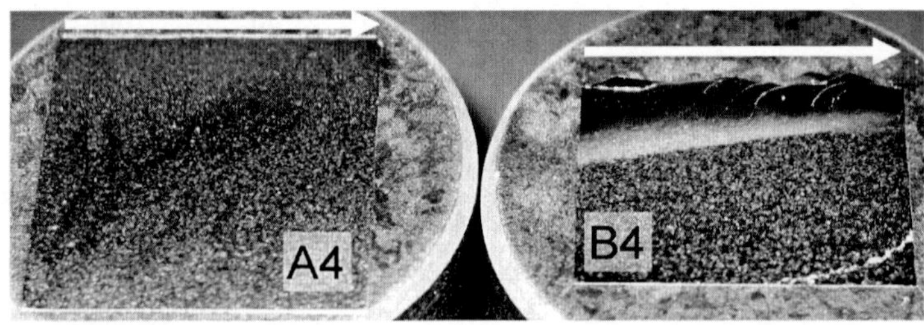

Fig. 4.4 Comparison of rail specimens (A4 and B4, sled velocity vector shown by arrows).

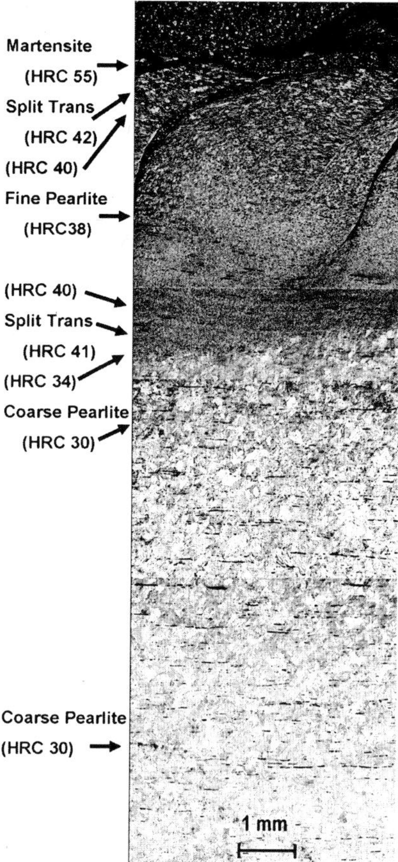

Fig. 4.5 Specimen B4—overall view (etched, sled travel is left to right).

The structural variation from the impact interface down into the body of the rail is very evident. In this preparation, the longitudinal "roll marks" from manufacturing process are clear. (In Fig. 4.5 they are the black horizontal striations and black spheroids through the thickness of the specimen.) It appears at this magnification that there are some clear zones of different microstructures. Also, there is a very clear demarcation between the affected zone and the unaffected rail below. This figure is marked with those microstructures identified, along with Rockwell hardness test C (HRC) values through the depth of the specimen. The justification and presentation of these microstructures appear in the remainder of this section.

The top of the specimen (i.e., the top of the gouge) was characterized by a significant amount of material flow, mixing (addressed later in this work), and evidence of rapid cooling—martensite. This is consistent with the previous results in [3]. Determination that this layer is martensite comes from the etch-resistant nature of the material, the measured HRC values (which correspond to the value for martensite appearing in literature [4, 5]), and an OIM test. The OIM evaluation identified the characteristic body-centered tetragonal (BCT) crystal structure, small grain size, and a lack of a preferred grain orientation (which implies rapid cooling, consistent with martensite formation [4, 5]). The carbon content of 1080 steel (0.8%) restricts residual austenite concentration to a maximum of 10%. The OIM analysis indicated the presence of austenite's face-centered cubic (FCC) crystal structure, but not in significant amounts. While this phase is present, its inclusion does not affect the conclusion of this analysis. Figure 4.6 shows the top of the specimen and the layer of martensite. The backscatter mode of the SEM is denoted as BSE (backscatter emitter).

Immediately below this top layer of martensite is a very distinct zone of microstructure that is a split transformation. A split transformation is a mixture of martensite and extremely fine pearlite that results from cooling from the austenizing temperature for 1080 steel (or eutectoid steel) of about 725°C. This kind of

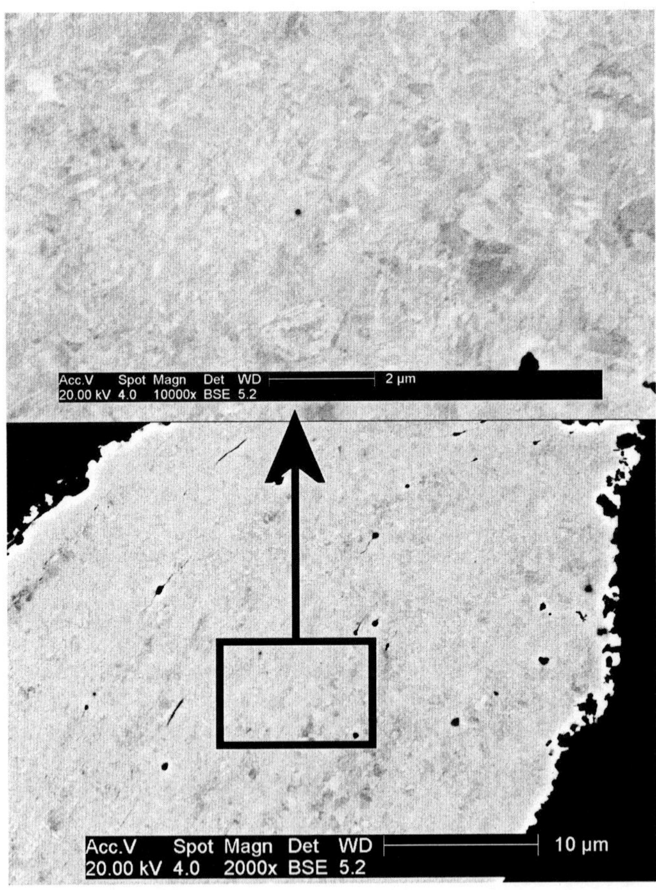

Fig. 4.6 Top of deformed specimen [as-polished, SEM, BSE (2,000× and 10,000×)].

microstructure is very similar to that reported in [5] for rapidly/continuously cooled eutectoid steel. The HRC value for this zone of microstructure is 42, which corresponds to a split transformation hardness value. Additionally, the OIM detected a mixture of BCT (martensite) and body-center cubic (BCC) crystal structure with a trace of FCC components (austenite). The BCC constituents represent ferrite (alpha phase), which along with cementite, form the fine pearlite. The grain sizes were small and showed no orientation preference, indicating a fast cooling rate [4, 5]. Figure 4.7 shows this microstructure, whereas Fig. 4.8 depicts the transition from this zone to one of fine pearlite. In these pictures, the lighter colored structures are martensitic, and the darker are pearlitic. A similar microstructure can be generated through an extended period of tempering (measured in many hours),

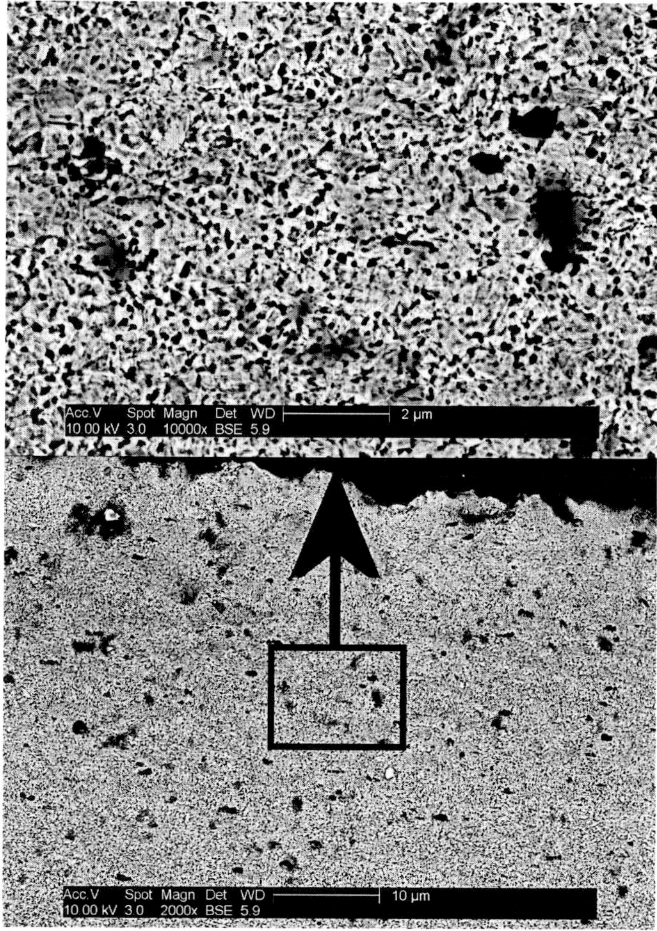

Fig. 4.7 Top of B4—split transformation [electropolished, SEM, BSE (2,000× and 10,000×)].

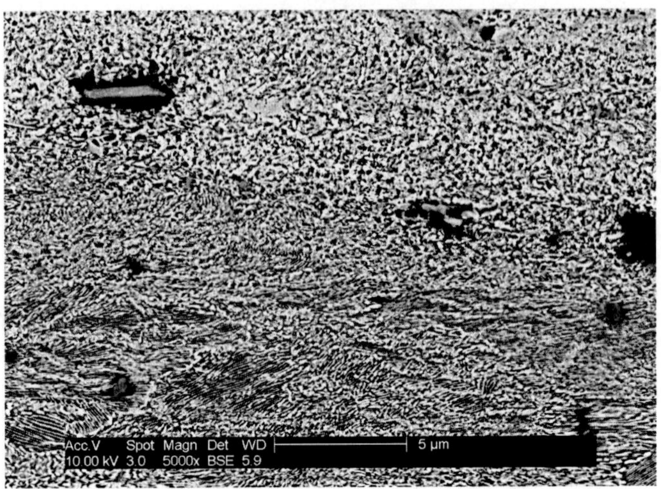

Fig. 4.8 Top of B4—split transformation transition [electropolished, SEM, BSE (5,000×)].

in which carbides precipitate out of the ferrite in a process known as spheriodization. However, we know from the HHSTT that these gouges cool to the touch in minutes because of rapid cooling. Additionally, the extreme energy of these hypervelocity impacts and evidence of melting on the railhead [3] allow us to conclude that the affected steel completely austenizes and that this microstructure is, in fact, a split transformation.

Examination, through the depth, below the top zone of split transformation revealed a zone characterized by fine pearlite. The HRC in this area also matches what has been reported in the literature for a fine pearlitic microstructure [4, 5]. The OIM examination yielded a predominant BCC crystal structure, which indicates pearlite and larger grain sizes. The grains had more of a preferred alignment, but were still indicating variability associated with fairly rapid cooling from austenite. Figure 4.9 illustrates this microstructure. Figure 4.10 is of one of the major shear bands in this area, with a layer of martensite along the fracture surface, and matches results reported in [3]. The presence of martensite was verified by hardness testing and optical metallurgy. This martensite forms as the material heated from plastic deformation rapidly cools when the material fractures.

Below the area of fine pearlite, at the edge of the zone that is visible in both Figs. 4.4 and 4.5, is another zone of split transformation. The HRC was measured as the higher value associated with the mixture of martensite and pearlite. The OIM again returned a zone of mixed crystalline structures and smaller grain sizes. Again, the grains showed no preferred orientation. This microstructure appears in Fig. 4.11.

Examination below this heat-affected zone revealed microstructure that transitioned immediately to coarse pearlite. This area yielded an HRC value consistent with typical coarse pearlite. The OIM analysis confirmed this with predominate BCC structure, large grains, and preferred orientations that characterize steel which

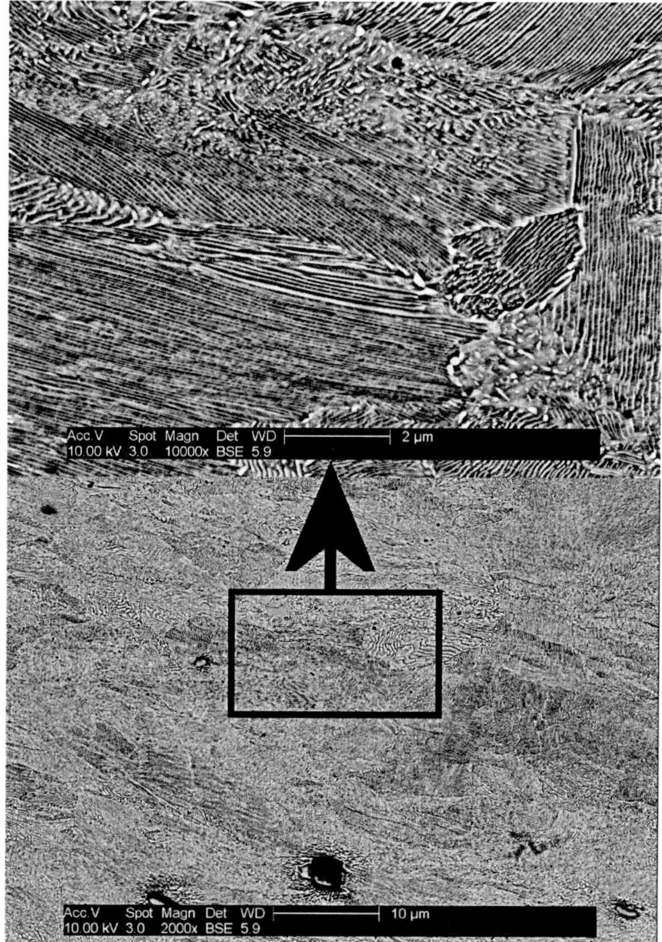

Fig. 4.9 Top of B4—fine pearlite [electropolished, SEM, BSE (2,000× and 10,000×)].

has cooled slowly. The coarse pearlite observed continues throughout the depth to the bottom of the specimen. Figure 4.12 shows this coarse pearlitic microstructure, and it is virtually identical to Fig. 4.13, which depicts the microstructure of specimen A4's undamaged microstructure. A reference OIM analysis performed on specimen A4 yielded similar results to those of the coarse pearlite in specimen B4.

This examination, through the depth, of the gouged rail has indicated various microstructures that match those in literature [4, 5]. The HRC hardness results and the OIM analysis match the microstructures identified by metallography. Unfortunately, a similar study with the sled shoes is not possible because of the destructive sled slow-down technique applied at the HHSTT.

To ascertain the material composition of the impact zone, EDS was performed on the damaged rail and on undamaged specimens of the rail (1080 steel) and

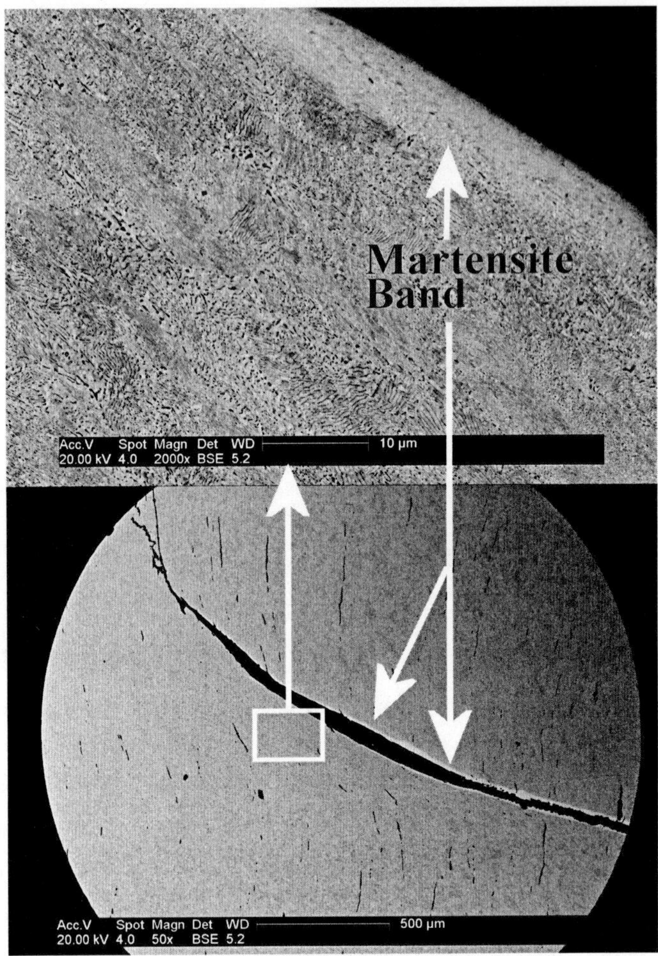

Fig. 4.10 Specimen B4—shear band [electropolished, SEM, BSE (50× and 2,000×)].

shoe (VascoMax 300). The damaged specimen matched the composition of the undamaged rail exactly with the exception of the area near the surface. At the top of the damaged rail, constituents such as nickel and cobalt (which is part of the composition of the shoe and not the rail steel) were found. This confirms a mixture of shoe and rail material that characterizes the gouging phenomenon.

IV. Analysis of Gouge Results

In the examination of this damaged rail section, it became clear that a large thermodynamic event was the mechanism that created a significant portion of the phenomenology present. There are microstructural changes in the deformed rail that indicate a large thermodynamic input into the event. That is, the material

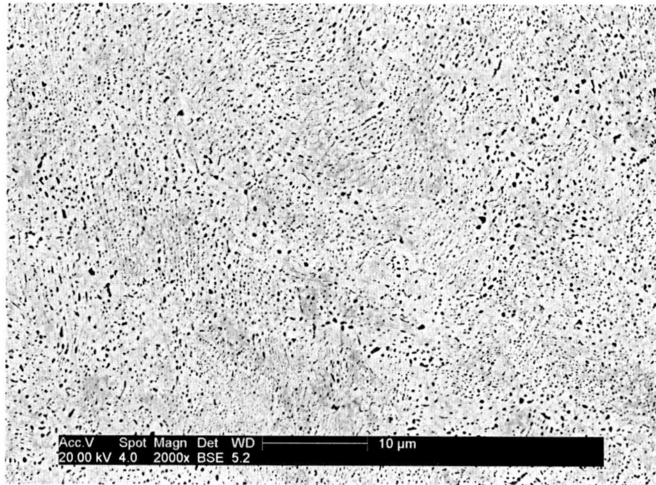

Fig. 4.11 Edge of heat-affected zone—split transformation [as-polished, SEM, BSE (2,000 x)].

damage must include a large temperature effect to account for the metallurgical results just presented.

From a macroscopic perspective, the shear bands resulting from catastrophic thermoplastic shear reported in [3] were verified. These shear bands match the character of the bands predicted by computer modeling of the impact event in

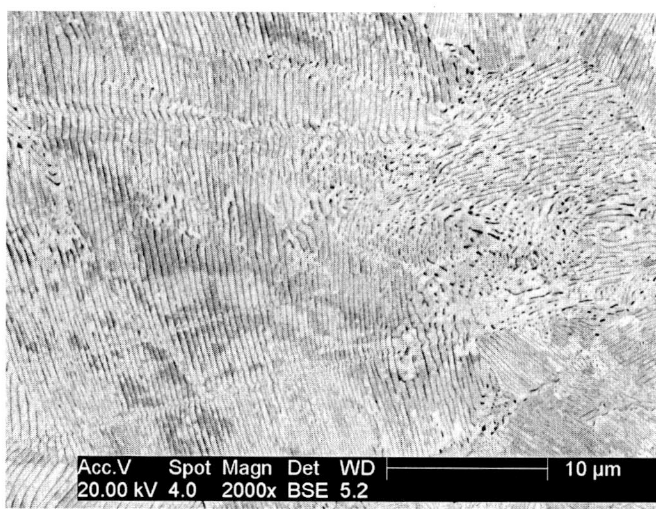

Fig. 4.12 Specimen B4—below heat-affected zone [as-polished, SEM, BSE (2,000 x)].

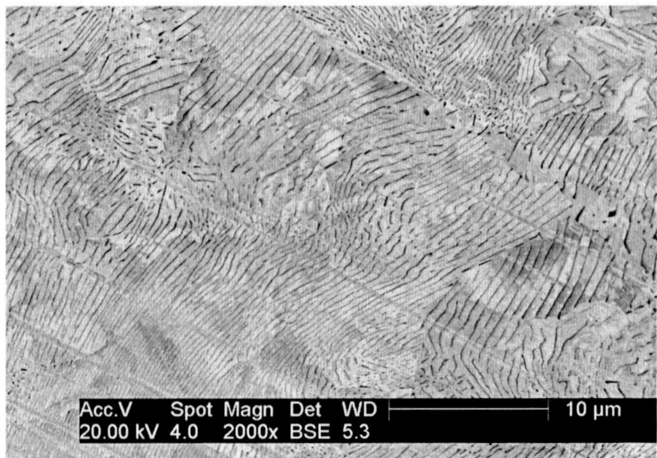

Fig. 4.13 Specimen A4 microstructure [as-polished, SEM, BSE (2,000×)].

[6–10]. The shear bands show evidence of martensite formation along the fracture surfaces consistent with thermoplastic activity followed by rapid quenching [3].

The experimentally observed microstructure suggests a thermodynamic history. The various types of final structure are the result of cooling from an austenizing temperature. In some applications, this quenching is done with oil or water baths. In others applications, the steel is allowed to air cool, which quenches the surface and the interior at different rates [4, 5]. The continuous (i.e., nonisothermal) cooling diagram for 1080 steel was obtained [5] and verified by the rail manufacturer for the HHSTT. This diagram, adapted from [5], appears as Fig. 4.14. Table 4.2

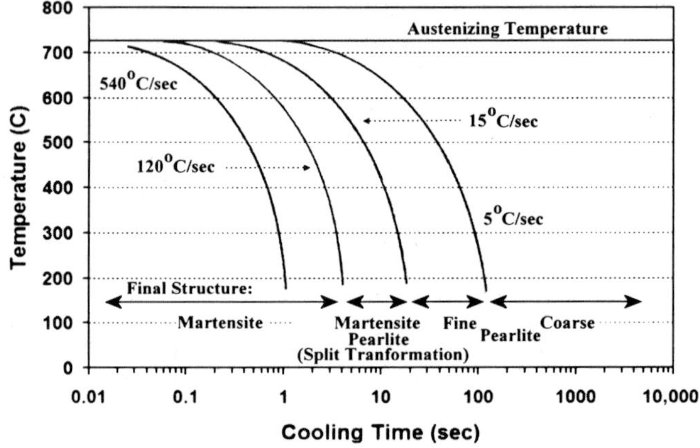

Fig. 4.14 Continuous cooling curve for 1080 steel (adapted from [5]).

Table 4.2 Summary of resulting microstructure from the continuous cooling of 1080 steel

Microstructure	Cooling rate, °C/s	Rockwell C value (HRC)
Martensite	120 or faster	50–65
Martensite/fine pearlite (split transformation)	15–120	40–50
Fine pearlite	5–15	35–40
Coarse pearlite	5 or slower	30–35

summarizes the microstructure resulting from various cooling rates when the steel starts above the austenizing temperature of 725°C.

Both the reports in [3] and our examination of the rail gouge indicate the surface of the rail had reached the melting temperature of 1480°C and therefore having the rail begin this process above the austenizing temperature seems very plausible. In addition, the models in [6–10] predict high temperatures into the rail at these depths.

Examining these events in terms of thermodynamics with continuous cooling, a description of the impact event that explains the microstructure revealed in experimentation can be proposed. The rail, undergoing plastic impact, heats above the melting temperature (1480°C) on the surface, and that material flows and partially sloughs off (seen in [3]). The rail heats above the austenizing temperature (725°C) down to a depth of approximately 7–10 mm below the original rail surface (or approximately 3 to 5 mm down into the remaining material below the gouge). Because this impact event only lasts in the neighborhood of 20 μs, the material below the heated zone is at the ambient temperature. That is, insufficient time (i.e., several seconds) is available for the conduction of the heat to occur, and no plastic deformation has occurred to generate viscoplastic temperature change.

When the deformation event has concluded (and the shoe has moved downrange), the rail immediately begins to cool. The heat is able to quench quickly into the air above the rail and into the steel below the austenized region, which explains the rapid cooling-related microstructure at the railhead and down at the lowest depth of the heat-affected zone. Rapid quenching into the air is aided by the turbulent airflow generated by the test sled traveling at speeds above 2 km/s. The ability of steel to quench rapidly both into the air and depth (when a part of a contiguous structure that was plastically deformed beyond the austenizing temperature) was noted in [3]. The area between these zones would cool more slowly.

The area below the heat-affected zone would not have heated during impact, but would heat as the thermal pulse travels into the depth of the rail. If this heat pulse were insufficient to increase the rail's temperature above the austenizing limit, then no appreciable microstructure change would occur. This section of the rail would be indistinguishable from the undamaged rail. Figure 4.15 illustrates this proposed thermal history.

A one-dimensional heat-conduction analysis can be used to validate the cooling rates suggested by the observed microstructure. Taking a one-dimensional slice through the depth, with the initial heat distribution noted in Fig. 4.15, creates

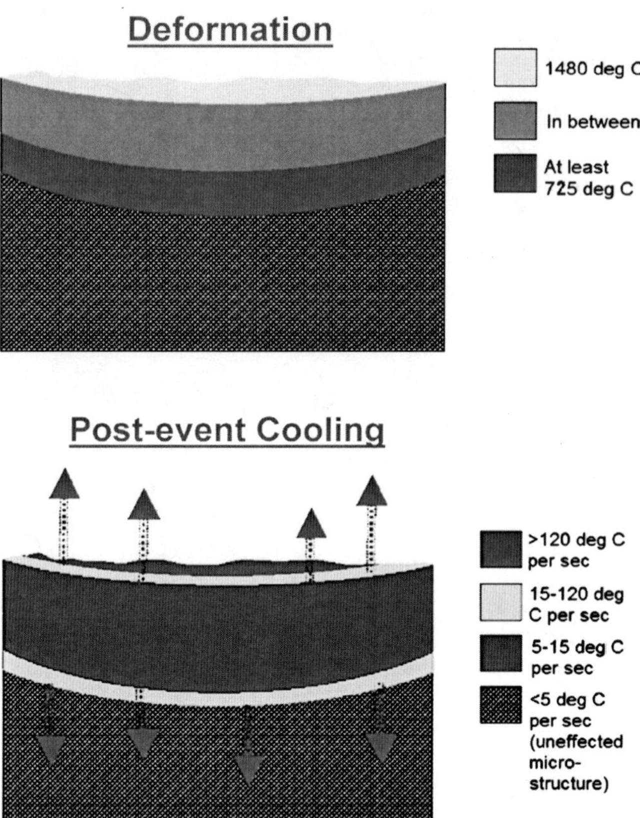

Fig. 4.15 Proposed thermal history of hypervelocity gouge.

the depiction appearing in Fig. 4.16. Note that x is defined as zero at the surface, down to ℓ. Because of the similarity of the heat profile and conduction across the gouge, this one-dimensional slice is a fairly accurate representation of the heat transfer path.

By applying Newton's Law of Cooling, we can arrive at the familiar heat-conduction relationship:

$$\frac{\partial T}{\partial t} = \frac{1}{\rho C_{\nu}} \nabla \cdot (K \nabla T) \tag{4.1}$$

where T is temperature, t is time, ρ is density, C_{ν} is specific heat at constant volume, and K is the thermal conductivity.

The one-dimensional form of this equation becomes

$$\frac{\partial T(x,t)}{\partial t} = \frac{1}{\rho C_{\nu}} \frac{\partial}{\partial x} \left(K \frac{\partial T}{\partial x} \right) \tag{4.2}$$

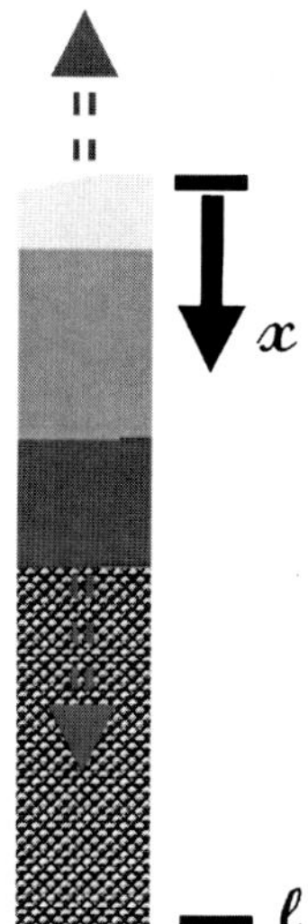

Fig. 4.16 One-dimensional slice of cooling gouge.

In Eq. (4.2) the x coordinate is defined in Fig. 4.16, and we note that K is a function of temperature. Typical solutions for this equation rely on establishing boundary conditions and holding K constant. As an example, let us consider a constant K and T to represent temperatures above ambient (300 K); therefore, the boundary conditions are zero at the rail surface and the bottom of our specimen.

Solving Eq. (4.2) in the standard manner, we arrive at the solution:

$$T(x,t) = \sum_{n=1}^{\infty} \left[\frac{2}{\ell} \int_0^{\ell} f(x) \sin\left(\frac{n\pi x}{\ell}\right) \mathrm{d}x \right] \cdot \sin\left(\frac{n\pi x}{\ell}\right) e^{-\lambda_n^2 t}$$

where

$$\lambda_n = \sqrt{\frac{K}{\rho C_v}} \cdot \frac{n\pi}{\ell} \tag{4.3}$$

where $f(x)$ represents the initial thermal profile estimated from knowing that the railhead was at the melt temperature (from [3] and experimental observations), and that the lowest depth of microstructure change must necessarily have been above the austenizing temperature. We can create an estimated temperature profile at time $t = 0$ by fitting a smooth curve through these known end conditions and forcing it to go rapidly to zero beyond the austenizing limit. This form matches those created by previous numerical simulation [6–10]. Our assumption that K is constant can be modified with experimental data concerning the thermal conductivity as a function of temperature. Table 4.3 summarizes this data.

Equation 4.2 was solved numerically, with K varying with temperature per Table 4.3. A linear fit was assumed between the discrete experimental points

Table 4.3 Summary of thermal conductivity of 1080 steel [11]

Temperature, °C	K [W/(m K)]
0	49.8
100	48.1
200	45.2
300	41.4
400	38.1
500	35.1
600	32.6
700	30.1
800	24.3
1000	26.8
1200	30.1

reported in Table 4.3. The results of this evaluation indicate that the estimated cooling rates needed to generate the presented microstructure are created. Figure 4.17 is a depiction of the entire range of dimensionality, from railhead to the bottom of the specimen. The various curves represent the temperature, through the specimen depth, at particular moments in time. The profile at time $t = 0$ was estimated as just described. Although the heat pulse moves down into the rail, the pulse does not move an austenizing temperature (700°C above ambient) lower into the depth.

If we examine the area of our concern, at temperature above the austenizing limit, we arrive at Fig. 4.18. More time points are represented in this figure to aid in our discussion.

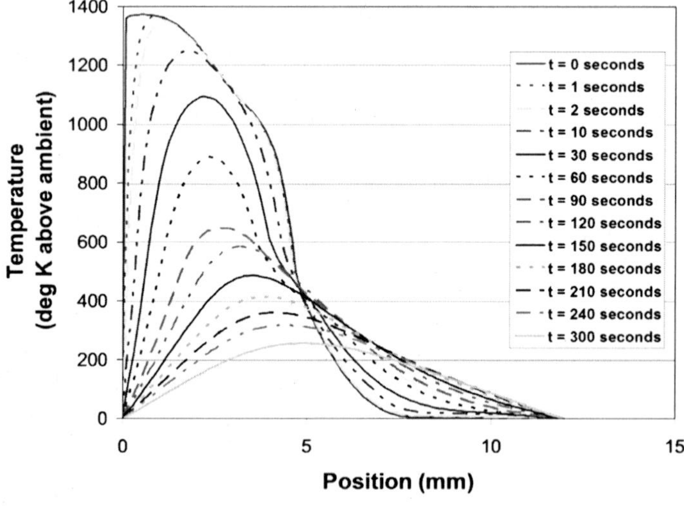

Fig. 4.17 1080 steel specimen cooling.

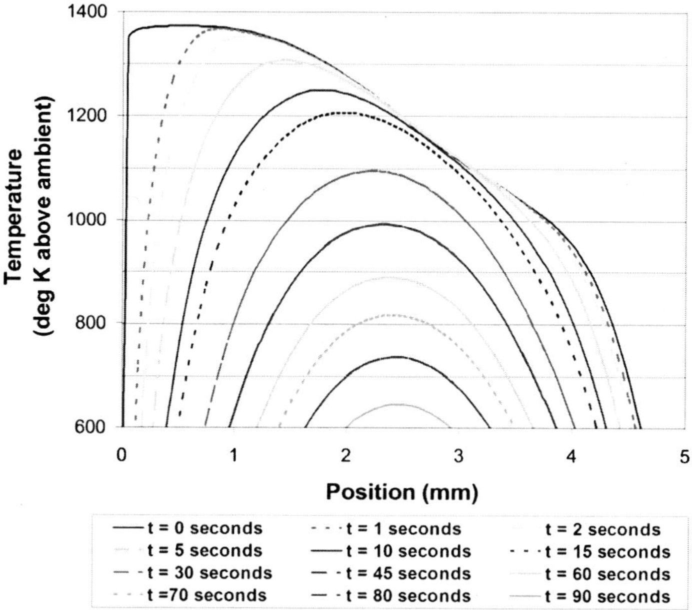

Fig. 4.18 Specimen cooling through austenizing temperature.

It is clear from the closer view that there is a thermal gradient established though the depth as the heat traveled to either the rail top or down into the rail depth. Figure 4.19 depicts the gradient as the temperature passes the critical austenizing limit. The figure also indicates the cooling rates required to generate the various microstructures and the cooling time for that particular rail depth position to cool below the austenizing temperature. Clearly, the thermal conductivity of 1080 steel is sufficient to create the microstructures identified from the HHSTT gouge. The layer of martensite at the gouge surface, the split transformation directly below the martensite, and the zone of fine pearlite are predicted exactly as we observed. Additionally, the split transformation within the depth of the specimen is predicted, along with the abrupt transition to unaffected microstructure.

An argument can be made that our boundary conditions do not account for the thermal conductivity to the air or are too aggressive. However, the observed microstructures indicate that this cooling did, in fact, take place. Additionally, the presented solution allows for a cooling rate that assumes the turbulent air of the sled passage could enforce an ambient air temperature. However, slower cooling rates that still create martensite on the rail surface are possible. The minimum cooling rates required to match our experimental results would only serve to adjust our model. That is, for martensite to be created on the gouge surface, the minimum thermal gradients required would match our model at any point down to 0.5 mm. Therefore, we could consider the gouge surface to be at 0.5 mm in our computational model, thereby making that amount of the model to be a thermal boundary layer, and still generate the microstructure experimentally observed throughout the depth of the specimen. The other boundary condition, within the depth, was

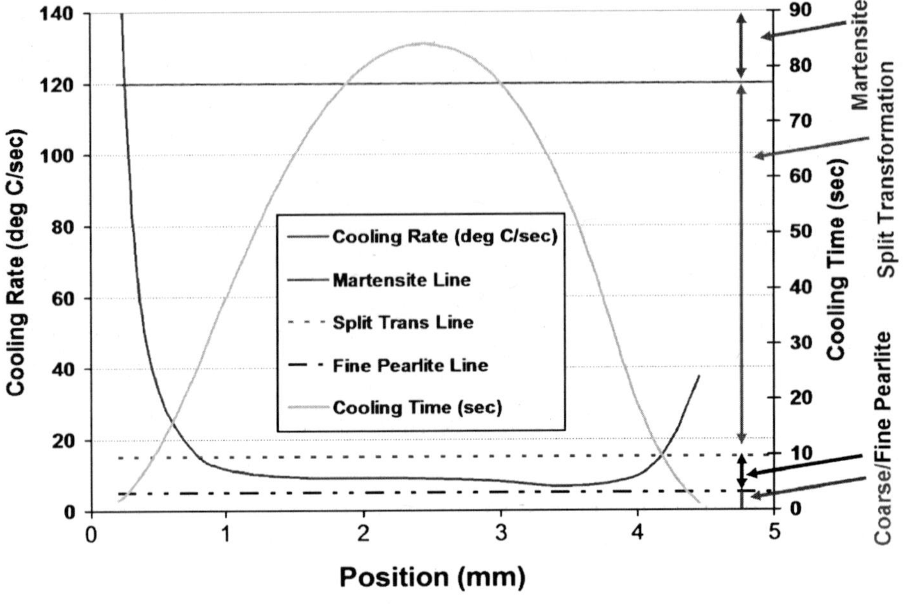

Fig. 4.19 Specimen cooling gradient and resulting microstructure.

varied to be at ambient temperature at distances further away from the surface, with little effect on the model. Therefore, our confidence that this one-dimensional model is sufficiently accurate to validate our experimental metallurgical results is very high.

An additional result of this analysis was to verify that the capability of conduction to transport temperature on the timescale of the impact event, as opposed to the cooling event, is very limited. Over the course of a typical impact, measured in the tens of microseconds, heat conduction results in only 1–2 K of temperature change. So although conduction during cooling is capable of causing the microstructure changes found in the experimental examination of the gouged rail, it is not a significant mechanism in the impact/deformation event. This fact was verified within previous CTH simulations of the impact event in [10].

V. Conclusions on Gouged Specimen Examination

A metallographic examination was conducted on a damaged rail having undergone hypervelocity gouging impact. The microstructure was shown to vary significantly from the specimen top down into the unaffected structure. A temperature history is suggested from the microstructural evidence. Material mixing in the region of the gouging, and the creation of shear bands from the heat generated by catastrophic thermoplastic shear, is confirmed in the micrographs. A one-dimensional heat-conduction model that exactly predicts the microstructure identified in the metallography is created. This concept is summarized in Fig. 4.20.

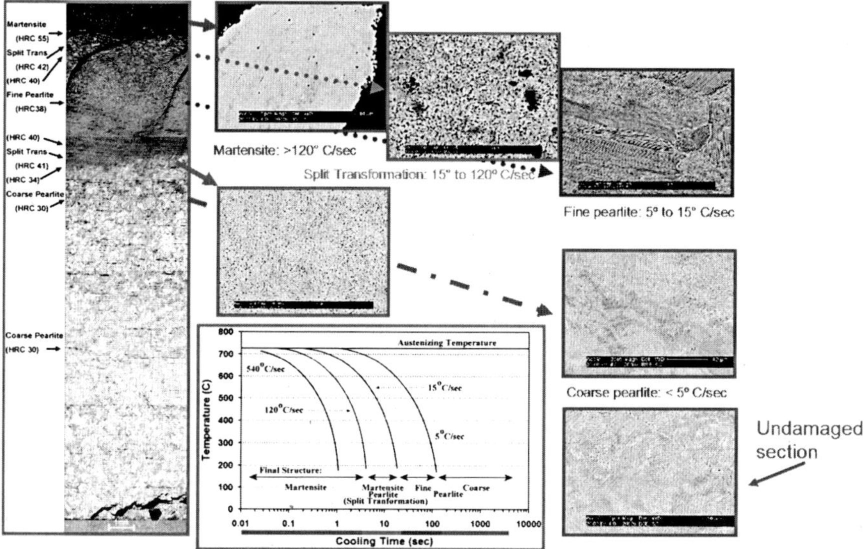

Fig. 4.20 Summary of gouged specimen examination.

The results of this analysis will serve to improve the numerical gouging model and further the effort to mitigate hypervelocity gouging at the HHSTT [12]. The results of this metallurgical study establish a thermal profile that can be matched against computer simulations of the gouging event. These code models of hypervelocity gouging can thereby be validated.

VI. Examination of Rail Condition

The discovery of such a large heat-affected zone and the associated thermal history that creates the observed microstructure prompted an investigation into other sections of rail from the HHSTT. In particular, two sections of rail that were not ever put into service were examined, as well as two sections that had been in service. One piece has no visible damage, and the other had a "scrape" under the flange where the shoe had contacted it as the sled had pitched or moved up. Table 4.4 summarizes the samples examined, their condition, and their nomenclature. All of these specimens were prepared per the "etched" procedure outlined in Table 4.1 and evaluated using bright-field microscopy. All rail samples were examined through the depth away from the surface into the bulk of the material.

Both of the virgin rail sections showed no thermally induced microstructure change, except at the rail head surface. A standard manufacturing process is to spray the rail heads with water during the cooling period, called "head" hardening, to create a thin, martensitic layer on the top surface [4]. This increases resistance to wear through the enhanced hardness of the top surface. Although this result was not expected by the HHSTT engineers (as the rails were not supposed to be head hardened), this small microstructural variance most likely has little impact on the gouging phenomenon. This judgment is based on the small dimensionality of the

Table 4.4 Summary of nongouged rail specimens

Type of rail	Coating	Damage	Sample location	Nomenclature
Unused	None	None visible	Rail head	E2K
Unused	Iron oxide	None visible	Rail head	E2KR
In service	None	None visible	Rail head	IS
In service	Iron oxide	None visible	Rail head	ISRT
In service	Iron oxide	Scrape	Under flange	ISRB

variation. Figures 4.21 and 4.22 are of these virgin rail sections. The black horizontal line in Fig. 4.21 denotes a break in the microscopy, and the microstructure far away from the rail head appears below that line. A similar technique could have been applied to Fig. 4.22, but was not necessary; the unaffected microstructure

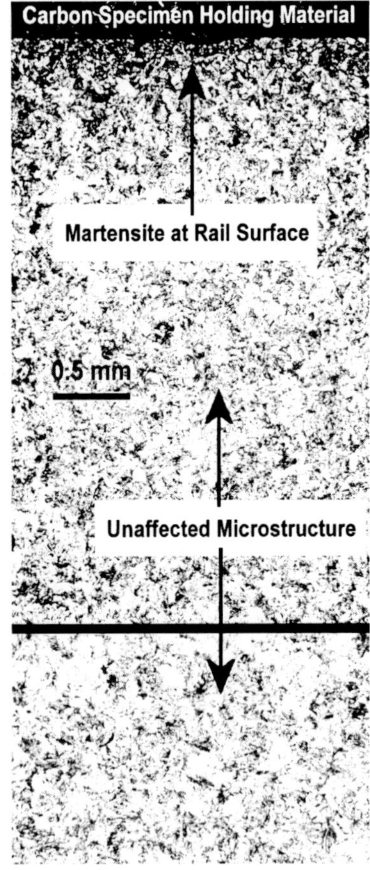

Fig. 4.21 Specimen "E2K" microstructure.

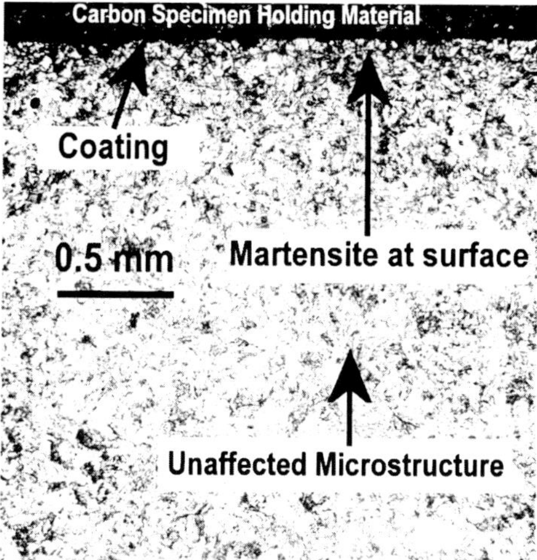

Fig. 4.22 Specimen "E2KR" microstructure.

(as with the preceding figure) looked identical as one proceeded through the depth away from the rail head.

The in-service rails were examined in the same fashion. However, it became immediately clear that microstructure changes similar to those appearing in Fig. 4.5 were present in these in-service rails.

Figure 4.23 is of an in-service rail that had no coating. The rail had no visible damage. A section of the rail head was examined, and an area of altered microstructure was found. This indicates that a microstructural change was caused by this rail section heating to above the austenizing temperature of 725°C and then cooling at a sufficiently rapid rate to generate the changes. The heating-conduction analysis earlier in this chapter indicates that insufficient time exists during shoe passage to conclude the heat pulse was a conduction-related event. Therefore, some small amount of plasticity or viscoelasticity, which accounts for the heat input into the depth of the rail while not resulting in obvious damage, might have occurred. This discovery also establishes a validation point to compare against a computational solution.

Another in-service rail was also examined. This particular rail section had a visible "scrape" on the underside of the flange. This was caused by the sled shoe rising and striking the underside of the rail head. The scrape was painted over again with iron-oxide coating, so that a visible evaluation was not possible. This scraped section was removed and examined through the "depth," which in this case was away from the ground and toward the top surface of the rail. Figure 4.24 depicts the microstructure observed. Once again, a section of altered microstructure exists near the rail surface, which indicates a thermal pulse that austenized the steel, followed by rapid cooling.

The scraped section of rail was also examined in an area away from the damage and where no visible damage was present. A section of the rail head was evaluated

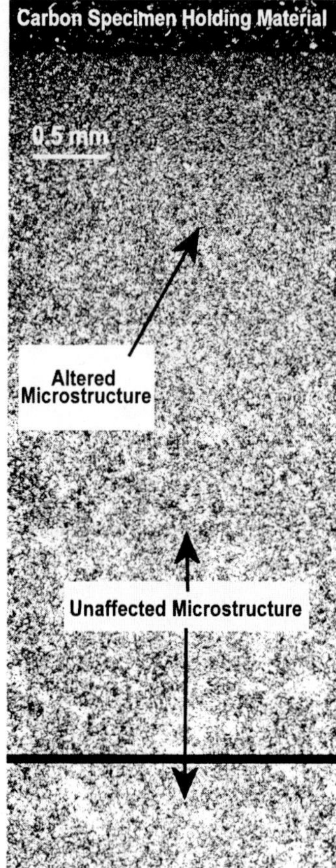

Fig. 4.23 Specimen "IS" microstructure.

and appears in Fig. 4.25. Although this section did not show visible damage, the altered microstructure is also present in this sample.

Earlier in this chapter, specimen A4 was also examined through the depth and used as a control for the evaluation of the hypervelocity gouge. The section of rail that specimen A4 came from was also a section of in-service and undamaged rail. Yet this rail showed no sign of microstructural change. Therefore, we can conclude that not all in-service rail has encountered heat pulses. However, some sections have suffered from this kind of thermal effect, most likely the result of small amount of plastic deformation.

VII. Conclusions

In this chapter, several rail sections were examined using metallurgical techniques in order to better understand gouging. One section had experienced a

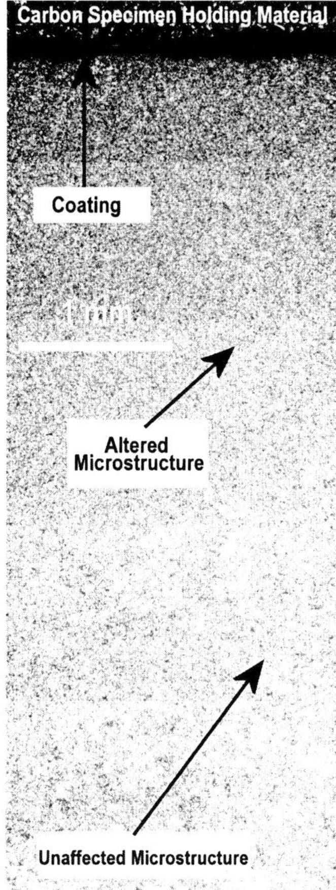

Fig. 4.24 Specimen "ISRB" microstructure.

hypervelocity gouge. Other sections were virgin specimens or had been in service on the HHSTT.

Based on the in-depth examination of the gouged rail, a thermodynamic history that explained the observed altered microstructure was developed and analytically verified. Material mixing, shear band development, and thermally induced phase changes were confirmed. Using the same approach, microstructural changes (very similar to those observed in the gouged rail) were discovered in the in-service rails. The timescales involved in these sled tests preclude the origin of these thermal pulses to be heat conduction from a heated shoe. Therefore, some relatively small amounts of plasticity or viscoelasticity must have caused the thermally induced microstructural changes observed.

These thermal profiles quantify, in some sense, the amount of heat that must have been generated by either a gouging impact, or a less severe material interaction, between the HHSTT sled and the rail. This information will serve as a validation

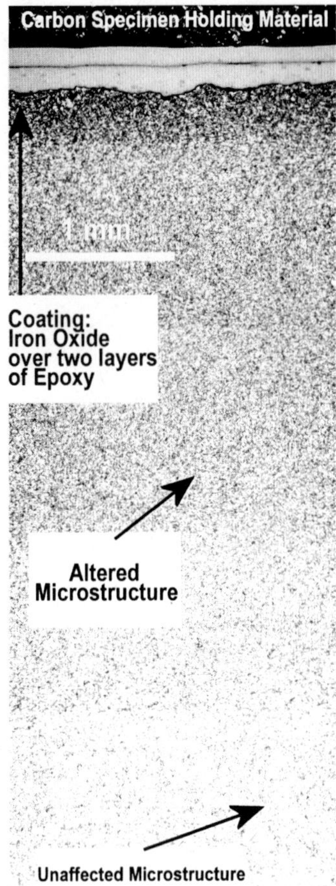

Fig. 4.25 Specimen "ISRT" microstructure.

point for the CTH simulations to replicate in the process of creating a usable impact model.

References

[1] Cinnamon, J. D., and Palazotto, A. N., "Metallographic Examination of Thermal Effects in Hypervelocity Gouging," *Proceedings of the 2005 ASME Pressure Vessels and Piping Division Conference*, ASME Paper PVP 2005-71613, July 2005.

[2] Cinnamon, J. D., and Palazotto, A. N., "Metallographic Examination and Validation of Thermal Effects in Hypervelocity Gouging," *ASME Journal of Pressure Vessel Technology*, Vol. 129, No. 1, 2007, pp. 133–141.

[3] Gerstle, F. P., Follansbee, P. S., Pearsall, G. W., and Shepard, M. L., "Thermoplastic Shear and Fracture of Steel During High-Velocity Sliding," *Wear*, Vol. 24, Issue 1, 1973, pp. 97–106.

[4] Davis, J. R. (ed.), *Metals Handbook, ASM International, Desk Edition*, 2nd ed., American Society for Microbiology, New York 1998.

[5] Smith, W. F., *Structure and Properties of Engineering Alloys*, McGraw–Hill, New York, 1981.

[6] Szmerekovsky, A. G., and Palazotto, A. N., "Structural Dynamics Considerations for a Hydrocode Analysis of Hypervelocity Test Sled Impacts," *AIAA Journal*, Vol. 44, No. 6, June 2006, pp. 1350–1359.

[7] Szmerekovsky, A. G., Palazotto, A. N., and Baker, W. P., "Scaling Numerical Models for Hypervelocity Test Sled Slipper-Rail Impacts," *International Journal of Impact Engineering*, Vol. 32, No. 6, 2006, pp. 928–946.

[8] Szmerekovsky, A. G., Palazotto, A. N., and Ernst, M. R., "Numerical Analysis for a Study of the Mitigation of Hypervelocity Gouging," AIAA Paper 2004-1922, April 2004.

[9] Szmerekovsky, A. G., "The Physical Understanding of the Use of Coatings to Mitigate Hypervelocity Gouging Considering Real Test Sled Dimensions AFIT/DS/ENY 04-06," Ph.D. Dissertation, Department of Aeronautics and Astronautics, Air Force Inst. of Technology, Wright-Patterson AFB, Dayton, OH, Sept. 2004.

[10] Szmerekovsky, A. G., Palazotto, A. N., and Cinnamon, J. D., "An Improved Study of Temperature Changes During Hypervelocity Sliding High Energy Impact," AIAA Paper 2006-2090, May 2006.

[11] Hoyt, S. L. (ed.), *Metals Properties*, ASME Handbook, 1st ed., McGraw–Hill, New York, 1954.

[12] Cinnamon, J. D., and Palazotto, A. N., "Refinement of a Hypervelocity Gouging Model for the Rocket Sled Test," *Proceedings of the 2005 ASME International Mechanical Engineering Congress and Exposition*, ASME Paper IMECE 2005-80004, ASME, New York, Nov. 2005.

Chapter 5

Constitutive Model Development

I. Introduction

AN ESSENTIAL element in the effort to accurately model hypervelocity gouging is the development of specific material constitutive models. In the HHSTT scenario, all previous work in the field has relied on material flow models that were not specific to VascoMax 300 and 1080 steel. In many cases, the material models available to the investigators significantly differed from these.

As discussed in Chapter 3, the material constitutive model is a very important element in the solution of hypervelocity impact [1]. Although the EOS tends to dominate the solution in some aspects, a significant portion of the solution depends on the material flow model. In this chapter, an extensive experimental study that investigates the flow characteristics of these specific materials is detailed.

II. Constitutive Model Overview

As noted in Chapter 3, the two constitutive model formulations that will be utilized are the Johnson–Cook and the Zerilli–Armstrong. Both of these approaches capture the major parameters that must be accounted for in a modern material flow model. That is, the dynamic yield strength of the material must be a function of strain, strain rate, and temperature. Therefore, an experimental approach that aims at resolving the constants in either of these constitutive models must vary these variables [see Eqs. (3.20) and (3.24)].

The strain-rate variation in the development of a constitutive model becomes the key in this effort. This is because the higher strain-rate tests are typically beyond the scope of most facilities and become expensive to conduct. Yet this higher strain-rate regime is exactly where these hypervelocity impacts occur [2]. Previous modeling efforts indicated that strain rates in the 10^3/s to 10^6/s range were common in these hypervelocity impacts [3–7]. Even though the EOS begins to dominate the solution at these higher pressures and strain rates, the material constitutive model makes a significant contribution to the solution [1].

To capture the maximum range of strain rates in the model development effort, two major tests are conducted. The first is the traditional split Hopkinson bar (SHB) test. This test can generate states of uniaxial strain rate up to 10^3/s. The second test is a flyer impact plate experiment, which creates uniaxial strain rates from 10^4/s to 10^6/s. These two tests span the entire range of interest and form the experimental basis upon which an accurate constitutive model can be formulated for VascoMax 300 and 1080 steel.

III. Split Hopkinson Bar Test

A series of SHB tests were conducted using VascoMax 300 and 1080 steel from the HHSTT. Because the HHSTT performs the heat treatment for the VascoMax 300 and to ensure we were testing the exact material from the field, the SHB specimens were prepared by the HHSTT machine shops.

A. SHB Test Background

A typical SHB test apparatus was used to test specimens of 1080 steel and VascoMax 300 [8–10]. The facility at the University of Dayton Research Institute (UDRI) was used. Figure 5.1 is an overall view of the test apparatus. The manner in which the specimens are heated in the test section of the apparatus is illustrated in Fig. 5.2. The test section, with a tested specimen, appears in Fig. 5.3.

The bars in this SHB apparatus are 0.5-in.-diam Inconel 718. The striker bar was capable of generating stress pulses that created strain rates in the test specimens of up to ~1500/s. The stress pulse is assumed to be

$$\sigma = \rho c_0 V_s \tag{5.1}$$

where ρ is the material density (7900 kg/m^3), c_0 is the material sound (elastic wave) speed, and V_s is the striker bar velocity. The striker bar velocity can be measured

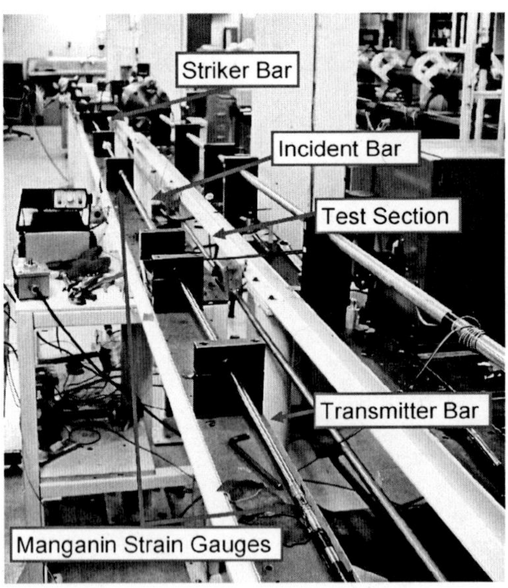

Fig. 5.1 UDRI SHB test apparatus.

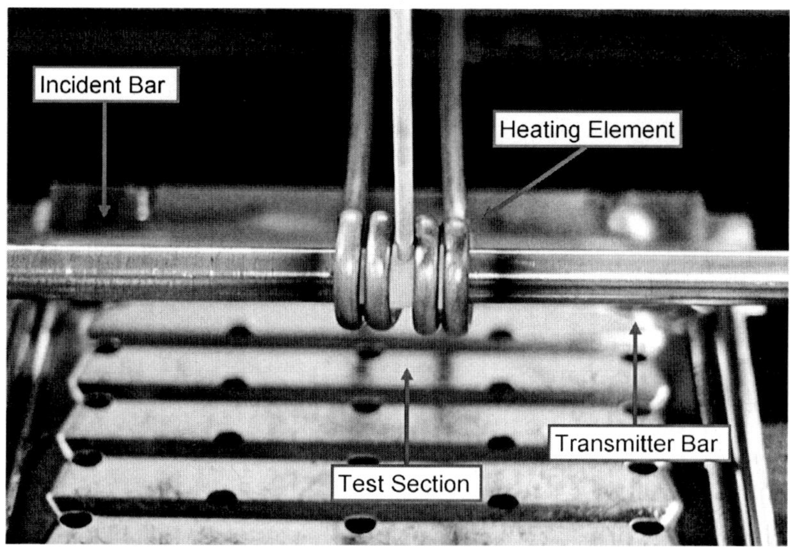

Fig. 5.2 SHB test apparatus heating element.

(using laser breaks), and the elastic wave speed can be found using (where E is the Inconel 718 bar elastic modulus)

$$c_0 = \sqrt{\frac{E}{\rho}} = \sqrt{\frac{195\,\text{GPa}}{7900\,\text{kg/m}^3}} = 4968\,\text{m/s} \tag{5.2}$$

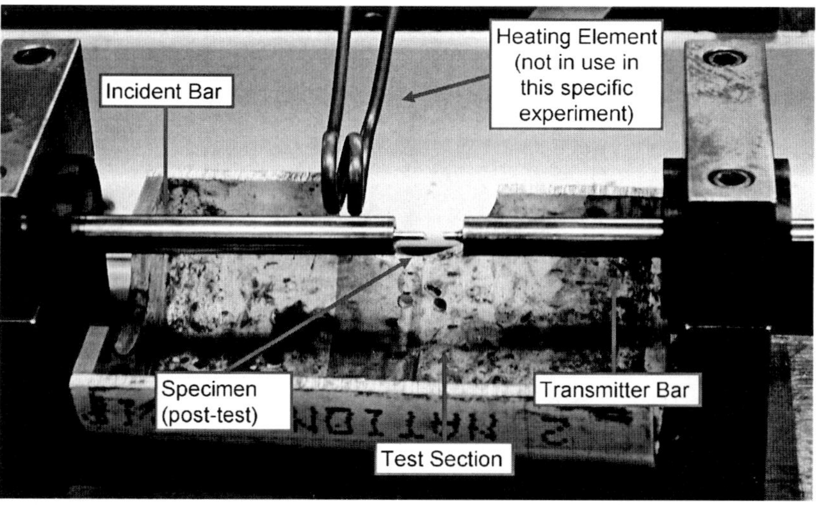

Fig. 5.3 SHB apparatus test section.

The created compressive stress pulse travels through the incident bar, through a collar surrounding the test specimen, to the end of the transmitter bar. The free end reflects the pulse back as a tensile wave that arrives back at the specimen (where the collar now has no effect because the collar is not attached to either bar). Figure 5.4 is a schematic of the test apparatus and shows this arriving tensile wave as ε_i, the incident strain wave. The incident wave is partially reflected as ε_r and transmitted as ε_t. The manganin strain gauges on the apparatus bars allow for the measurement of these strain pulses.

Following the theory developed in [1, 11–15], the values of specimen strain rate and stress can be computed from these strain measurements.

The displacements of the ends of the specimen in Fig. 5.4 can be expressed in Eq. (5.3), where $\varepsilon = \partial u/\partial x$ and $\sigma = E\varepsilon$, where u is displacement, σ is uniaxial stress, and ε is uniaxial strain. The average strain in the specimen can be found from Eq. (5.4), where L is the length of the specimen test section. The forces P at the ends of the specimen can be computed in Eq. (5.5) from noting that $\sigma = E\varepsilon = P/A$, where E is the test material elastic modulus and A is the Hopkinson bar cross-sectional area.

$$\left\{ \begin{aligned} u_1 &= \int_0^t c_0 \varepsilon_1 \,\mathrm{d}t = c_0 \int_0^t (\varepsilon_i - \varepsilon_r)\,\mathrm{d}t \\ u_2 &= \int_0^t c_0 \varepsilon_2 \,\mathrm{d}t = c_0 \int_0^t \varepsilon_t \,\mathrm{d}t \end{aligned} \right\} \tag{5.3}$$

$$\varepsilon_s = \frac{u_1 - u_2}{L} = \frac{c_0}{L} \int_0^t (\varepsilon_i - \varepsilon_r - \varepsilon_t)\,\mathrm{d}t \tag{5.4}$$

$$\left\{ \begin{aligned} P_1 &= EA\varepsilon_1 = EA(\varepsilon_i + \varepsilon_r) \\ P_2 &= EA\varepsilon_2 = EA\varepsilon_t \end{aligned} \right\} \tag{5.5}$$

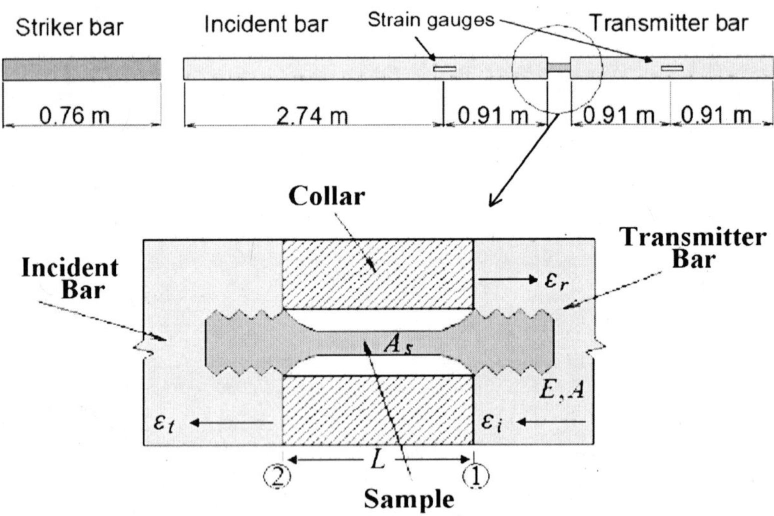

Fig. 5.4 SHB test apparatus schematic.

Assuming the forces are the same at both ends of the specimen, Eq. (5.5) implies that $\varepsilon_i + \varepsilon_r = \varepsilon_t$, and therefore from Eq. (5.4)

$$\varepsilon_s = \frac{c_0}{L}\int_0^t (\varepsilon_t - \varepsilon_r - \varepsilon_r - \varepsilon_t)\,\mathrm{d}t = -\frac{2c_0}{L}\int_0^t \varepsilon_r\,\mathrm{d}t \tag{5.6}$$

which is the specimen strain. This is available from the strain-gauge measurements of ε_r. The force at the specimen ends must equal the force in the bars, which requires

$$\sigma_s = \frac{A}{A_s}\sigma_b = \frac{A}{A_s}E\varepsilon_t \tag{5.7}$$

where σ_b is the stress in the bar and A_s is the gauge cross-sectional area of the test specimen. The specimen strain rate is obtained from Eq. (5.6) (by differentiating with respect to time) as

$$\dot{\varepsilon}_s = -\frac{2c_0}{L}\varepsilon_r \tag{5.8}$$

With these relationships a set of material data (stress and strain) can be gathered at varying strain rates and temperature. From these data, a constitutive model can be created.

B. SHB Test Results

A series of SHB tests were conducted on material specimens machined at the HHSTT to be identical to those materials in use in the field. The specimens were machined to the specifications appearing in Fig. 5.5.

The 1080 steel test results were typical of a strain-hardening material. Figure 5.6 shows a typical stress–strain curve generated by the SHB. The full report of the results is available in [16]. Consistent with the assumptions within the SHB

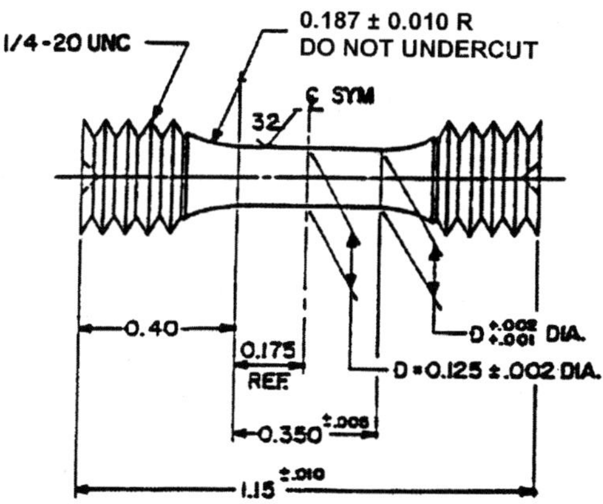

Fig. 5.5 . Specifications for the SHB specimens (units are inches).

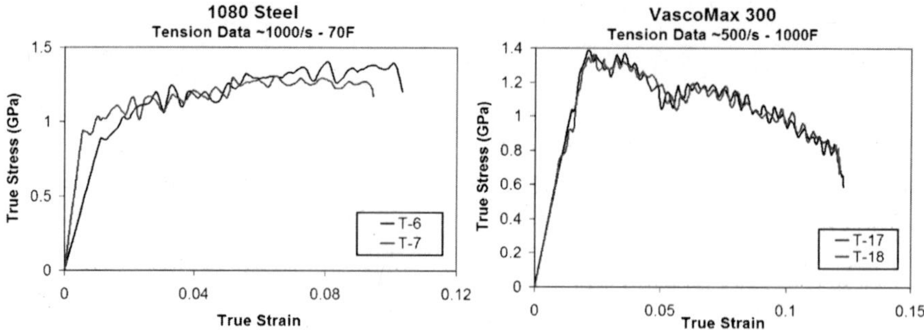

Fig. 5.6 Comparison of stress–strain behavior of subject materials under tensile SHB test.

relationships, the 1080 steel specimens showed no measurable necking in the specimens. Therefore, the material stress can be computed, via Eq. (5.7), from the measured strain and the curves can thereby be constructed. Table 5.1 summarizes these tests results. The quasi-static tests (at a strain rate of ~1/s) were conducted using the SHB specimens and a standard quasi-static pull test machine.

VascoMax 300, on the other hand, did not behave as a typical strain-hardening material. Figure 5.6 shows a typical VascoMax 300 stress–strain profile. It exhibits little strain hardening before the material begins to fail. Additionally, the specimens experienced significant necking during the testing process. Table 5.2 summarizes the VascoMax test results.

Early in this research process, a technique that adjusted the results of these tests for the necking observed in the VascoMax 300 specimens was applied [8–10]. It was thought that the necking process was part of the strain-hardening process and that the material was capable of carrying more stress as a result. However, it became clear during additional testing that VascoMax 300, in its current state

Table 5.1 Summary of 1080 steel SHB results

Test no.	Test temp, °F	Strain rate, s^{-1}	Flow stress at $\varepsilon \sim 0.06$, GPa
Q1, Q2	70	~1	1.048
3, 4	70	~500	1.22
11, 12	300	~500	1.01
16, 17	500	~500	0.89
18, 20	750	~500	1.00
6, 7	70	~1000	1.27
13, 14	300	~1000	0.88
23, 24	500	~1000	0.75
22, 31	750	~1000	0.99
25, 26	70	~1500	1.18
33, 34	300	~1500	1.26
27, 28	500	~1500	1.12
36, 38	750	~1500	0.82

Table 5.2 Summary of VascoMax 300 SHB results

Test no.	Test temp, °F	Strain rate, s^{-1}	Flow stress at $\varepsilon \sim 0.01$, GPa
Q1, Q2	70	~1	2.15
3, 4	70	~500	2.2
15, 16	500	~500	1.78
19, 20	750	~500	1.75
17, 18	1000	~500	1.4
1, 2	70	~1000	2.18
9, 28	500	~1000	1.88
10, 11	750	~1000	1.63
12, 13	1000	~1000	1.33
6, 7	70	~1500	2.38
21, 23	500	~1500	1.95
24, 25	750	~1500	1.68
26, 27	1000	~1500	1.65

of heat treatment used at the HHSTT, has very little strain capability prior to failure. Therefore, the necking process is part of the failure of the material, and the strain-hardening portion is a small area prior to the drop in the true stress curve.

C. SHB-Based Constitutive Model Development

At this point, it is possible to develop a constitutive model based on the strain-rate data from 1/s to 1500/s. The rationale for developing the models at this point, as opposed to constructing them considering the flyer plate tests, is that one must sacrifice accuracy in matching the midrange strain rates of the SHB tests to match the high strain-rate flyer plates. Later in this chapter, it will become clear that perfectly matching all of the experimental data is not possible. Therefore, because constitutive models for these materials (VascoMax 300 and 1080 steel) have not been presented in the literature, the mid-strain-rate range models will be computed.

The procedure for reducing the SHB data and determining the constants for the material flow models appears in another work [16]. Essentially, by considering each portion of the flow model separately, a systematic approach can be made in determining the constants. For instance, evaluating the SHB experiments at room temperature and at 1/s strain rate, the first set of Johnson–Cook constants can be determined. Next, one can hold the strain constant and examine the SHB tests at varying strain rates to determine the next constant. Finally, the last constant is determined from examining the SHB tests are various temperatures. Finding the Zerilli–Armstrong constants is performed in a similar manner. In addition, the typical values for other metals like those considered can be used as a guide in the iteration process.

Using this approach, the model constants for both materials can be developed for VascoMax 300 and 1080 steel, based on the experimental data from the SHB tests. The typical manner in which the constitutive model is presented in the literature is through the use of an effective flow stress vs effective strain-rate diagram at a particular value of strain. In this way, the critical elements of strain-rate dependency

Table 5.3 Summary of physical properties and model constants from SHB

Property/constant	1080 steel	VascoMax 300
E, GPa	202.8	180.7
ν	0.27	0.283
T_{melt}, K	1670	1685
ρ, kg/m^3	7800	8000
JC: A, GPa	0.525	2.17
JC: B, GPa	3.59	0.124
JC: C	0.029	0.0046
JC: m	0.7525	0.95
JC: n	0.6677	0.3737
ZA: A, GPa	0.75	1.0
ZA: c_1, GPa	2.5	2.5
ZA: c_2, GPa	0	0
ZA: c_3, eV^{-1}	110.0	40.0
ZA: c_4, eV^{-1}	5.5	2.0
ZA: c_5, GPa	0.266	0.266
ZA: n	0.289	0.289

and thermal softening are presented. Table 5.3 summarizes the physical properties and the Johnson–Cook (denoted by JC) and Zerilli–Armstrong (denoted by ZA) constants derived from the SHB data, where E is the elastic modulus, ν is poisson's ratio, T_{melt} is the melting temperature in Kelvin, ρ is material density, and eV are units of electron volts. A conversion between electron volts and degrees Kelvin can be performed, if necessary, by noting that $1\,\text{eV} = 11,605\,\text{K}$.

Figures 5.7 and 5.8 are the resulting diagrams from the SHB data. In these diagrams, the "Exper." notation refers to the SHB experiments at specific temperatures in dgrees Fahrenheit, while JC and ZA refer to the Johnson–Cook and Zerilli–Armstrong model predictions, respectively.

The validation of these models will be discussed in detail in the next chapter. These midrange strain-rate constitutive models would be extremely useful in a lower-energy impact simulation or other scenario in which the strain rates are in the neighborhood of 10^3/s or lower. For the computational simulation of the shoe/rail impact, however, a higher strain-rate regime model is necessary. Therefore, high strain-rate tests were conducted.

Another element that the SHB tests can provide is an estimation of the ultimate pressure/stress of the material. This value is the maximum stress measured from the stress–strain diagrams before the materials begin to fail and the stress drops. For VascoMax 300 and 1080 steel, the values are approximately 2.5 and 2.0 GPa, respectively.

IV. Flyer Impact Plate Experiments

To generate data to extend previously developed flow models, a higher strain-rate uniaxial test is required. The maximum strain rate that can be generated in

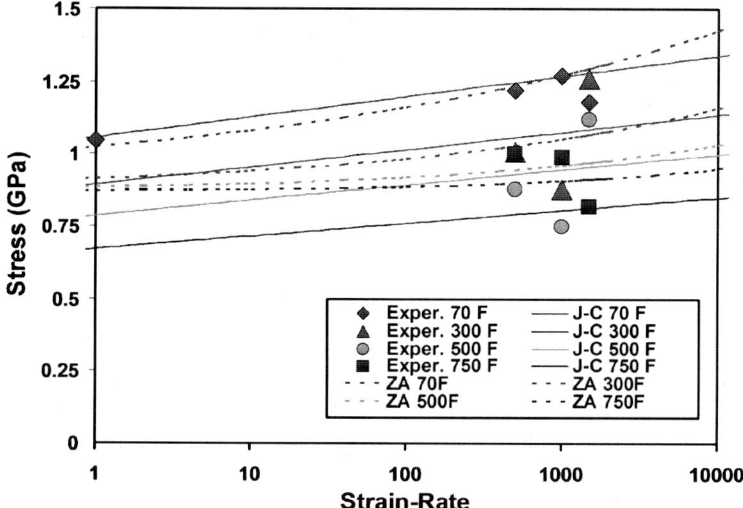

Fig. 5.7 Effective flow stress vs strain rate, 1080 steel, SHB.

the SHB scenario is on the order of 10^3/s. The magnitudes required to extend the constitutive model are in the 10^4/s–10^5/s range. The type of test that is typically employed is the flyer plate impact experiment [17]. These high velocity impact tests can provide stress measurements with respect to time for a given impact [18].

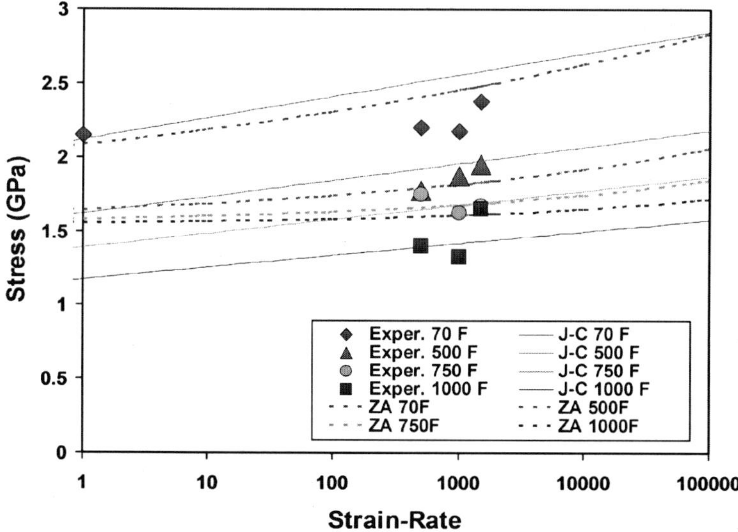

Fig. 5.8 Effective flow stress vs strain rate, VascoMax 300, SHB.

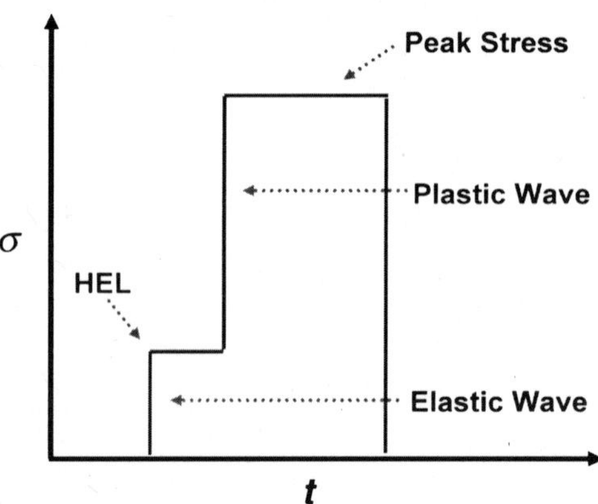

Fig. 5.9 Idealized stress vs time plot for a uniaxial planar impact.

These stress curves are characterized by an elastic precursor wave, followed by the plastic deformation wave, as illustrated in Fig. 5.9. The magnitude of the elastic precursor wave is known as the Hugoniot elastic limit (HEL). This value provides one of the accepted estimates for the dynamic yield (flow) stress of the material. The flyer plate experiments can be performed in such a manner as to yield both the HEL and the peak stresses at given impact velocities.

The elastic precursor wave travels at the material sound speed [see Eq. (2.1)]. The plastic deformation wave moves at a slower speed behind the elastic precursor. At a particular point in the material, the peak stress will be the summation of the plastic and elastic deformation waves, until the elastic release wave returns from the far-field boundary. The reason that the HEL is used to estimate the flow stress at a given strain rate is that uniaxial strain simulations of the impact event will rely on the material constitutive model to adjust the magnitude of the HEL—and thereby the total peak stress.

Therefore, these flyer plate experiments provide unique information in the formulation of a material constitutive model. The Johnson–Cook or Zerilli–Armstrong models serve to establish the magnitude of the HEL, while the material EOS governs the material behavior at the stresses experienced in the plastic deformation wave. Therefore, these experiments require a uniaxial strain code simulation of the impact conditions, with the capability to adjust the material constitutive models and the equation of state. Both of these elements could be adjusted to refine the material models to match the experimental results.

For our experiments, the HHSTT facility once again manufactured the test specimens for the flyer impact plate experiment.

A. Flyer Plate Experiment Background

To extend the previously determined material models for VascoMax 300 and 1080 steel to the higher strain-rate regime, a series of flyer plate experiments was conducted [19]. The tests were conducted at the UDRI. The test facility appears in Fig. 5.10.

Fig. 5.10 UDRI flyer plate testing facility.

One of the two test configurations performed appears in Fig. 5.11. This first set of tests (one each for the specific materials) was conducted with the goal of recording the HEL. The flyer plate (3 mm thickness) was shot against a 6-mm-thick target of the same material, with a manganin stress gauge attached to the back of the plate, held by 12 mm of polymethyl methacrylate (PMMA). Because of the sufficient target thickness, the elastic precursor wave was able to separate from the plastic wave of deformation. This allowed measurement of both the HEL and the peak stress wave that followed it.

Two additional experiments were conducted for each material in the second test configuration. For these tests, the target was composed of 2 mm of material, a stress gauge, and 12 mm of additional material. These shots were performed to record the stress wave within the material as it passes. The rationale for performing these additional tests was to establish an experiment to record the in-material stress waves for model validation at lower impact velocities and therefore lower peak stresses. This geometry did not result in the detection of the HEL (because of a

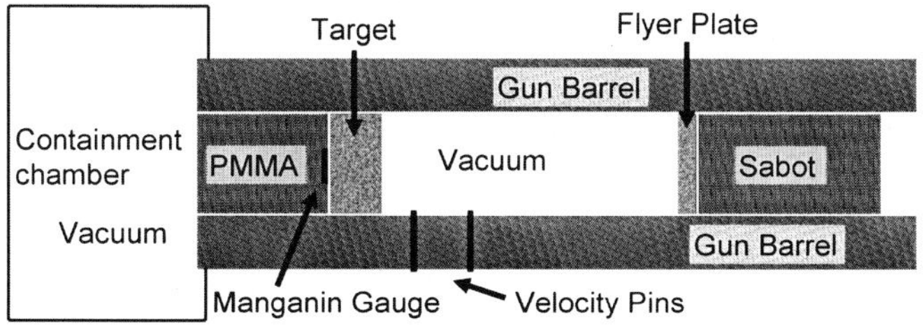

Fig. 5.11 Schematic of flyer plate experimental tests 7-1878 and 7-1879.

Fig. 5.12 UDRI flyer plate and target.

thickness that was not sufficient for the elastic wave to separate from the plastic wave) and is illustrated in Fig. 5.12. In both types of tests, the manganin stress gauge was electrically insulated from the steel by a Mylar sheet of 0.025-mm thickness. Table 5.4 summarizes the test conditions.

The stress-gauge measurements were taken for each of the cases, as well as impact velocity information from the velocity pins. The stress measurement was recorded as a function of time as the elastic and plastic waves traveled through the gauge.

B. Flyer Plate-Based Constitutive Model Development

To formulate a constitutive model for these materials, a one-dimensional wave code is necessary. This is because the HEL and the peak stresses need to be simulated as a function of time and compared to the experimental results. The hydrocode, CTH, was chosen to simulate these impacts through a one-dimensional model. Because CTH will be used in hypervelocity impact simulations, using this particular code and its EOS formulation was essential to building an accurate constitutive model.

Table 5.4 Summary of flyer plate tests

Test	Flyer	Velocity, m/s	Target
7-1874	3 mm VascoMax 300	685	2 mm + gauge + 12 mm VascoMax 300
7-1875	3 mm 1080 steel	669	2 mm + gauge + 12 mm 1080 steel
7-1876	3 mm VascoMax 300	450	2 mm + gauge + 12 mm VascoMax 300
7-1877	3 mm 1080 steel	437	2 mm + gauge + 12 mm 1080 steel
7-1878	3 mm VascoMax 300	891	6 mm VascoMax 300 + gauge + 12 mm PMMA
7-1879	3 mm 1080 steel	891	6 mm 1080 steel + gauge + 12 mm PMMA

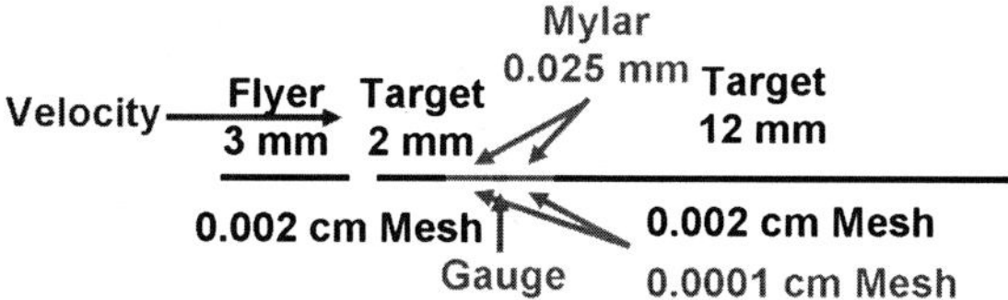

Fig. 5.13 CTH model for flyer test configuration one.

Based on previous research, a mesh convergence study has indicated that a mesh size of 0.002 cm is where the CTH solution converges for these impact simulations and which is at the edge of where continuum mechanics is considered to end for these metals [6]. Therefore, a mesh size was chosen to be 0.002 cm throughout the model of the flyer impact scenarios, except where the Mylar sheet is against the gauge. In those cases, the mesh size chosen was 0.0001 cm. Figures 5.13 and 5.14 depict the CTH models used for these two flyer test geometries.

Tests 7-1878 and 7-1879 were simulated first. The modification of the constitutive model affected the magnitude of the HEL prediction, as well as the peak stress. Figures 5.15 and 5.16 depict the CTH simulation of the stress pulse as it travels through the gauge for Test 7-1878 (VascoMax 300) based on the baseline JC and ZA models developed from the SHB tests (see Table 5.3). The experimental data are plotted on the same figure, as well as annotations for the HEL, predicted HEL, and where the material failed during testing. The HEL and the peak stress predictions are fairly close, but could be improved. The difference between simulation and experiment is far more dramatic in Figs. 5.17 and 5.18. These figures depict the baseline JC and ZA predictions for the stress in the 1080 steel tests. The HEL is significantly underpredicted, as is the peak stress.

By modifying the constitutive models in the higher strain-rate regime, the CTH predictions for the stress wave could be adjusted. Although the EOS dominated the solution, up to a 10% modification could be made to the peak stress by allowing for higher strengths at strain rates of 10^5/s (which is the level computed by CTH as the strain rate of the flyer plate tests). As expected, these modifications to the

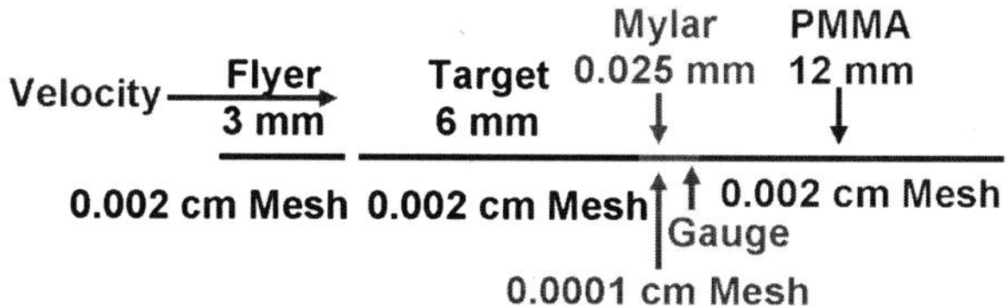

Fig. 5.14 CTH model for flyer test configuration two.

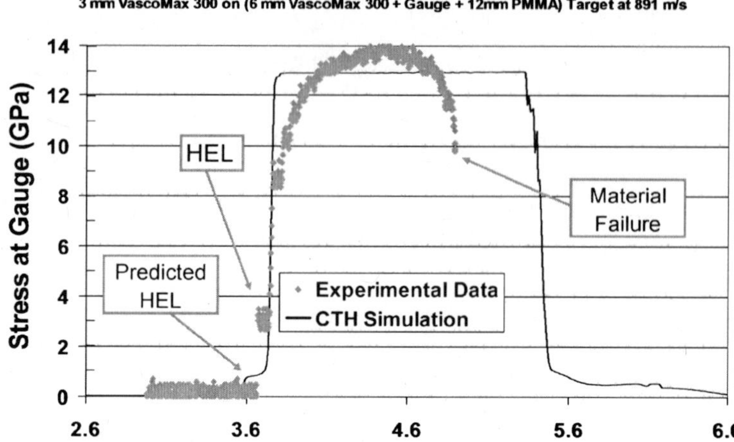

Fig. 5.15 Baseline JC, CTH simulation, test 1878.

constitutive model were necessary to make the HEL prediction match the flyer plate experiments. The previous models, developed from the SHB tests, were underpredicting the flow stress at the strain-rate levels experienced in the flyer plate impacts. By requiring these adjustments, the flyer plate experiments were

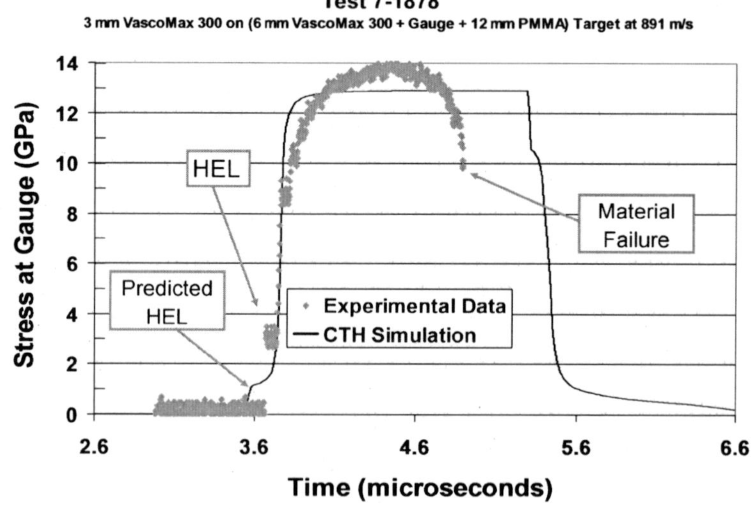

Fig. 5.16 Baseline ZA, CTH simulation, test 1878.

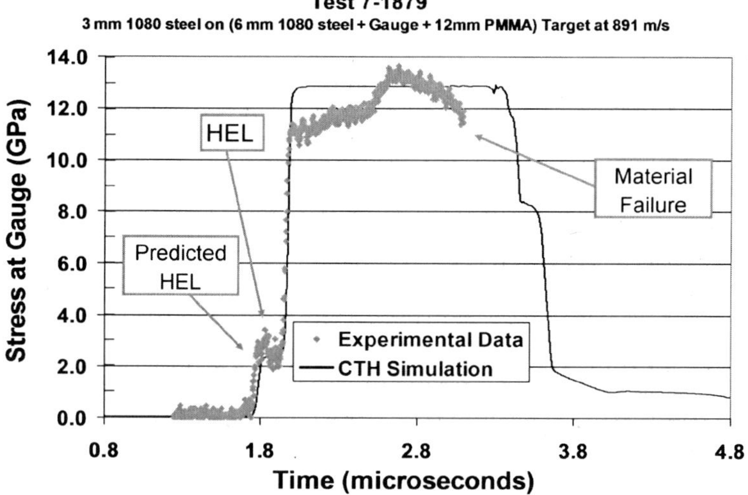

Fig. 5.17 Baseline JC, CTH simulation, test 1879.

effectively bridging the gap between the lower strain-rate SHB tests and the very high strain-rate impacts dominated by the EOS calculations.

An iteration process was undertaken to adjust the parameters of the Johnson–Cook and Zerilli–Armstrong models to find a best match between the CTH simulations and the experimental flyer plate tests. That is, the constants of these

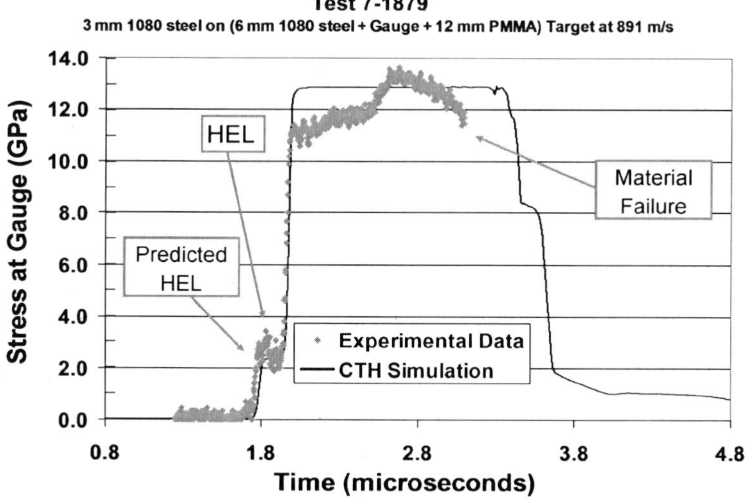

Fig. 5.18 Baseline ZA, CTH simulation, test 1879.

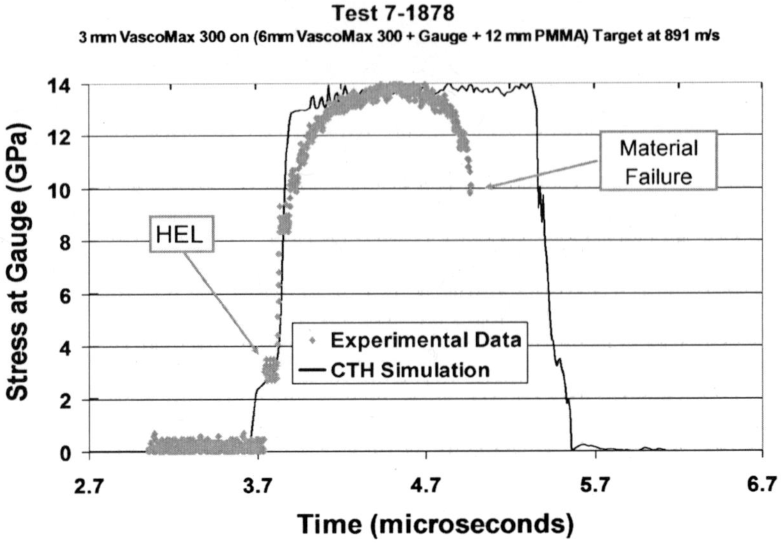

Fig. 5.19 Best-fit JC, CTH simulation, test 1878.

models were adjusted, and then the CTH code was used to simulate the flyer plate tests, and a comparison was made between the measured stress and the CTH predicted stress. The model constants were iterated to create a "best-fit" match for all three tests available for each material (1878, 1874, and 1876 for VascoMax 300 and 1879, 1875, and 1877 for 1080 steel).

Initially, it appeared as if the Johnson–Cook model (which had been employed previously [8–10, 20, 21]) might continue to be sufficient as the constitutive model for these materials. However, because of the linear relationship between effective stress and strain rate, the higher stress estimates for 10^5/s strain rate required to match the flyer plate tests made the model overestimate the SHB regime data (illustrated later in this section). Because of this, the greater flexibility of the Zerilli–Armstrong model became important. It was possible to construct a Zerilli–Armstrong model that came close to matching the SHB data and still made the fit to the flyer plate data possible.

Figures 5.19 and 5.20 illustrate the JC and ZA "best-fit" CTH simulations for VascoMax 300 (test 1878) that matched all of the flyer plate data as much as possible. They depict the CTH simulation of the shock stress as a function of time for the models developed to match all of the flyer plate experiments. On the figures, the experimental stress-gauge data are depicted on the simulation. In addition, where the material failed during the experiment is annotated on the figure. Figures 5.21 and 5.22 illustrate the same information for 1080 steel (test 1879).

In comparing Figs. 5.19 to 5.22, both the Johnson–Cook and Zerilli–Armstrong formulations appear capable of matching the experiment tests fairly well. However, the difference between these formulations is far more evident when we examine the effective stress vs strain-rate plots for these models.

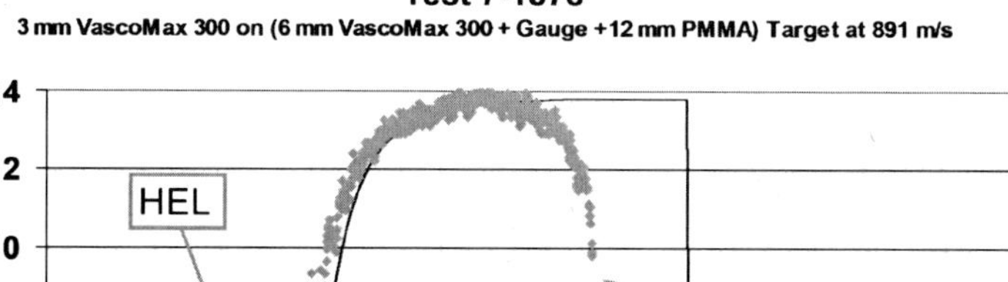

Fig. 5.20 Best-fit ZA, CTH simulation, test 1878.

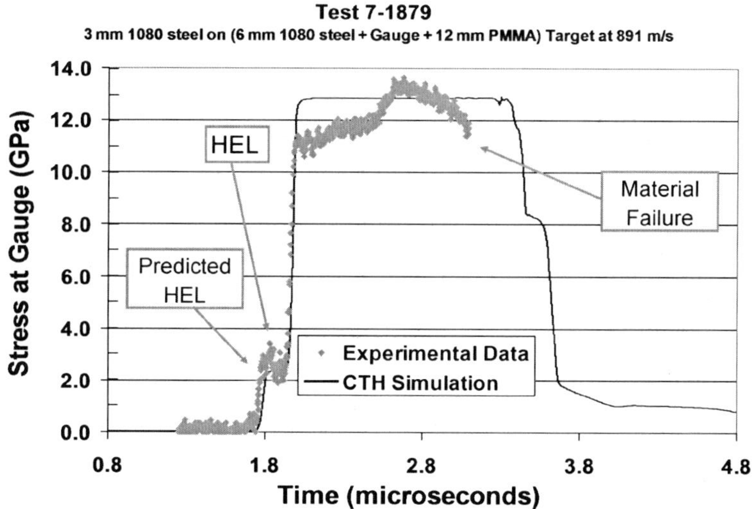

Fig. 5.21 Best-fit JC, CTH simulation, test 1879.

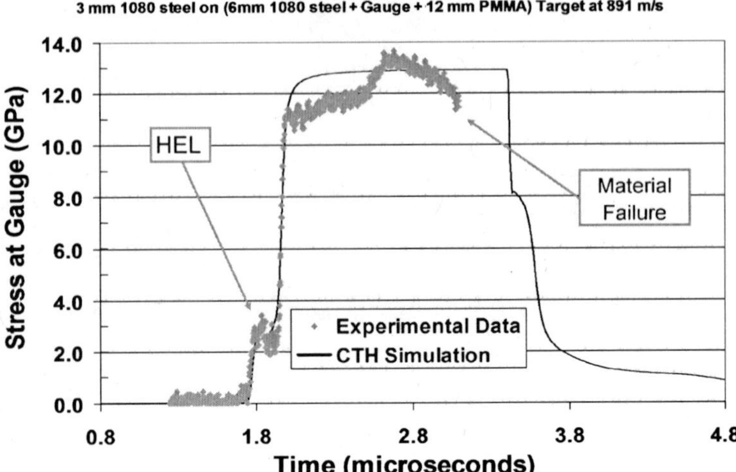

Fig. 5.22 Best-fit ZA, CTH simulation, test 1879.

Figures 5.23 and 5.24 illustrate the resulting constitutive relationships developed from matching the flyer plate data. To match the measured stress wave, the Johnson–Cook formulation must abandon the midrange strain-rate data from the SHB tests to a greater extent. Based on the requirement to develop a constitutive model that can estimate material flow stress across the entire strain-rate range, it

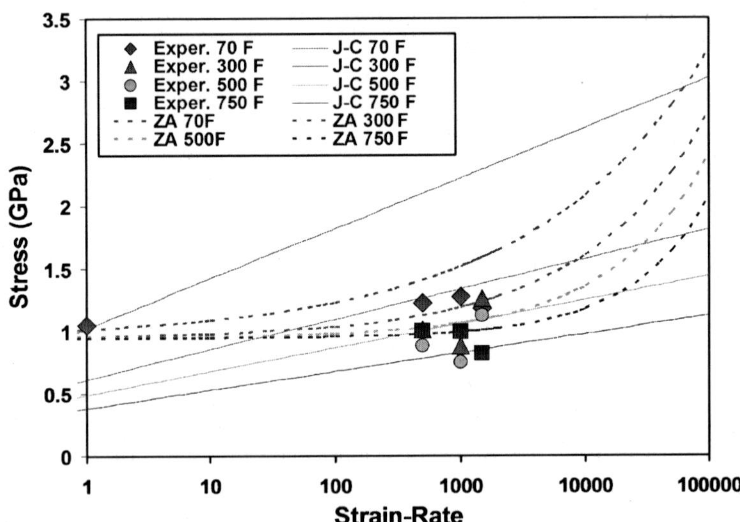

Fig. 5.23 Effective flow stress vs strain rate, 1080 steel.

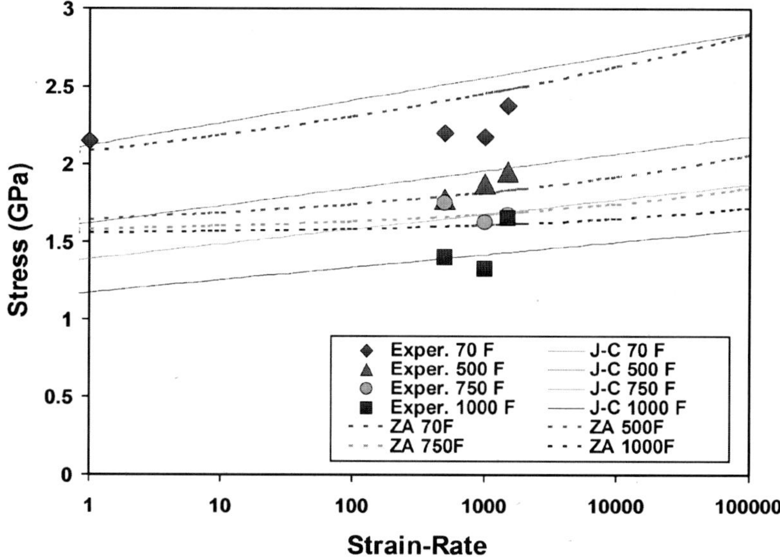

Fig. 5.24 Effective flow stress vs strain rate, VascoMax 300.

is clear that the Zerilli–Armstrong formulation is superior in its ability to maintain better matches to both sets of experimental data.

Table 5.5 summarizes the constants for both models that were determined to match the flyer plate stress data. Because of the better fit to the SHB data, the

Table 5.5 Summary of physical properties and model constants from flyer plate tests

Property/constant	1080 steel	VascoMax 300
E, GPa	202.8	180.7
ν	0.27	0.283
T_{melt}, K	1670	1685
ρ, kg/m^3	7800	8000
JC: A, GPa	0.7	2.1
JC: B, GPa	3.6	0.124
JC: C	0.17	0.03
JC: m	0.25	0.8
JC: n	0.6	0.3737
ZA: A, GPa	0.825	1.42
ZA: c_1, GPa	4.0	4.0
ZA: c_2, GPa	0	0
ZA: c_3, eV^{-1}	160.0	79.0
ZA: c_4, eV^{-1}	12.0	3.0
ZA: c_5, GPa	0.266	0.266
ZA: n	0.289	0.289

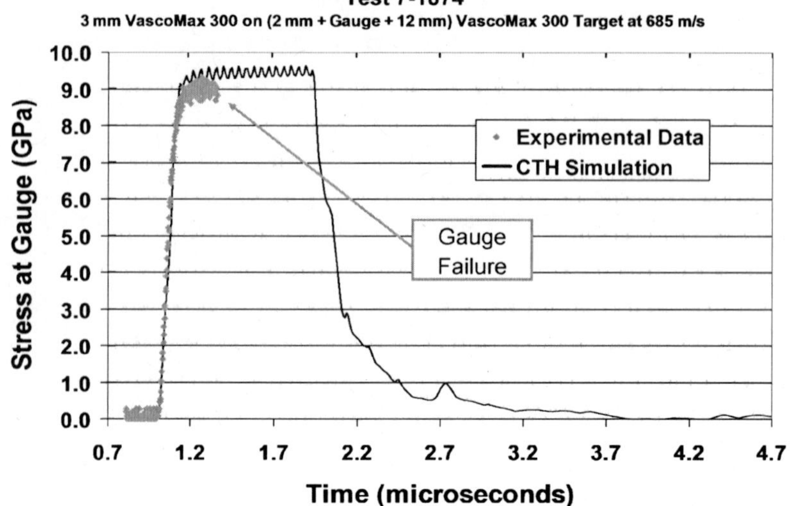

Fig. 5.25 CTH simulation, test 1874.

Zerilli–Armstrong model was chosen as the optimum model for the remainder of the investigation.

Figures 5.25–5.28 illustrate the fit that the Zerilli–Armstrong model and the EOS in CTH are able to make to the experimental flyer plate tests 1874–1877,

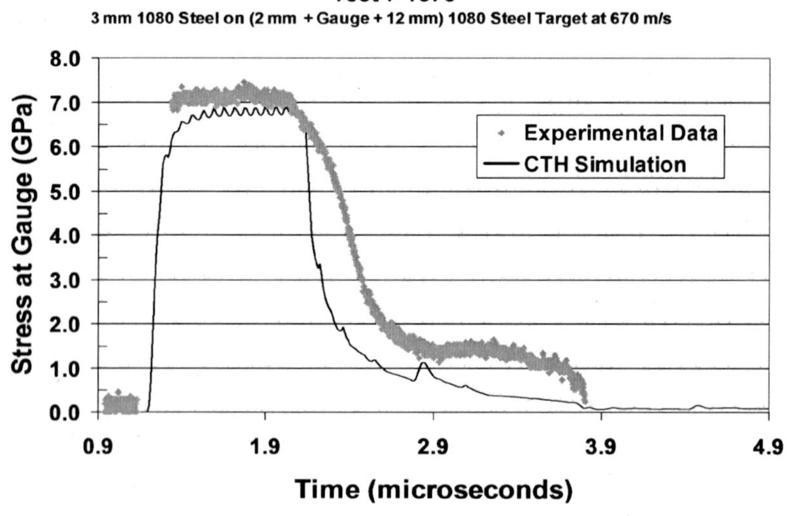

Fig. 5.26 CTH simulation, test 1875.

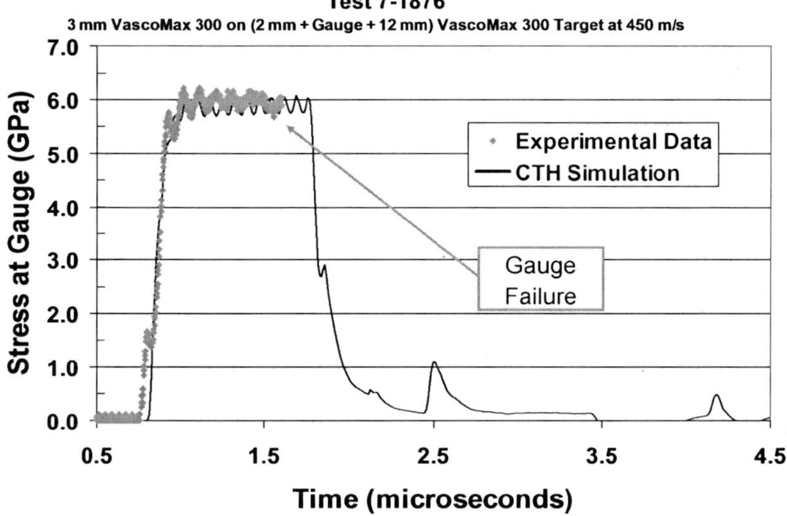

Fig. 5.27 CTH simulation, test 1876.

which were performed in the second configuration (with the "embedded" stress gauge). The stress wave was measured within the material, as opposed to at the rear of the target. Additionally, these tests were conducted at a lower velocity to provide varying peak stresses to match the models against. On these figures, an annotation is made where the gauge failed, which occurred in the VascoMax 300 experiments. This was further experimental evidence, akin to the SHB results, that VascoMax 300 has very little strain capability prior to failure.

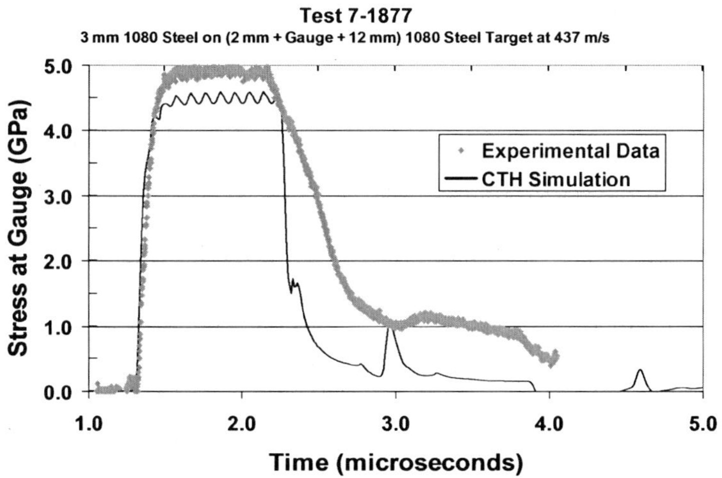

Fig. 5.28 CTH simulation, test 1877.

Based on the results, a very accurate constitutive model that creates a close match to the flyer plate impact test data and the previously performed SHB tests has been developed. Although the JC model developed from the flyer plate experiments tended to produce similarly good results to those appearing in Figs. 5.25–5.28, the poor fit to the SHB data was a concern. This comparison necessitates the use of the Zerilli–Armstrong model for further simulations in CTH of these materials undergoing dynamic deformation.

V. Summary

By conducting a series of experiments on the materials present in the HHSTT gouging problem (VascoMax 300 and 1080 steel), the material constitutive models were developed. In an attempt to create flow models that are experimentally based across the entire strain-rate range from 1/s to 10^5/s, the Zerilli–Armstrong formulation is chosen to continue the effort to model hypervelocity gouging. With these models, the next logical step is to validate them. This was accomplished by comparing CTH simulations to experimentation and ensuring that CTH replicates them reasonably well.

References

[1] Zukas, J. A., Nicholas, T., Swift, H. F., Greszczuk, L. B., and Curran, D. R., *Impact Dynamics*, Krieger, Malabar, FL, 1992.

[2] Nicholas, T., and Rajendran, A. M., "Material Characterization at High Strain Rates," *High Velocity Impact Dynamics*, Wiley, New York, 1990, pp. 127–296.

[3] Szmerekovsky, A. G. and Palazotto, A. N., "Structural Dynamics Considerations for a Hydrocode Analysis of Hypervelocity Test Sted Impacts," *AIAA Journal*, Vol. 44, No. 6, June 2006, pp. 1350–1359.

[4] Szmerekovsky, A. G., Palazotto, A. N., and Baker, W. P., "Scaling Numerical Models for Hypervelocity Test Sled Slipper-Rail Impacts," *International Journal of Impact Engineering*, Vol. 32, No. 6, 2006, pp. 928–946.

[5] Szmerekovsky, A. G., Palazotto, A. N., and Ernst, M. R., "Numerical Analysis for a Study of the Mitigation of Hypervelocity Gouging," AIAA Paper 2004-1922, April 2004.

[6] Szmerekovsky, A. G., "The Physical Understanding of the Use of Coatings to Mitigate Hypervelocity Gouging Considering Real Test Sled Dimensions AFIT/DS/ENY 04-06," Ph.D. Dissertation, Air Force Inst. of Technology, Wright-Patterson AFB, Dayton, OH, Sept. 2004.

[7] Szmerekovsky, A. G., Palazotto, A. N., and Cinnamon, J. D., "An Improved Study of Temperature Changes During Hypervelocity Sliding High Energy Impact," AIAA Paper 2006-2090, May 2006.

[8] Cinnamon, J. D., and Palazotto, A. N., "Refinement of a Hypervelocity Model for the Rocket Sled Test," *Proceedings of the 2005 ASME International Mechanical Engineering Congress and Exposition*, ASME Paper IMECE 2005-80004, ASME, New York, Nov. 2005.

[9] Cinnamon, J. D., Palazotto, A. N., and Kennan, Z., "Material Characterization and Development of a Constitutive Relationship for Hypervelocity Impact of 1080 Steel and Vascomax 300," *International Journal of Impact Engineering*, Vol. 33, No. 12, 2006, pp. 180–189.

[10] Cinnamon, J. D., Palazotto, A. N., Kennan, Z., Brar, N. S., and Bajaj, D., "Johnson-Cook Strength Model Constants for Vascomax 300 and 1080 Steels," *Proceedings of the 14th APS Topical Conference on Shock Compression of Condensed Matter*, Vol. 845, American Physical Society, College Park, MD, 2006, pp. 709–712.

[11] Gray, G. T., III, "High Strain-Rate Testing of Materials: The Split Hopkinson Bar," *Methods in Materials Research*, Wiley, New York, 1987.

[12] Lindholm, U. S., "Some Experiments with the Split Hopkinson Pressure Bar," *Journal of the Mechanics of Physical Solids*, Vol. 12, 1964, pp. 317–335.

[13] Mear, M. E. and Hutchinson, J. W., "Influence of Yield Surface Curvature on Flow Localization in Dilatant Plasticity," *Mechanics of Materials*, Vol. 4, Issues 3–4, 1985, pp. 395–412.

[14] Meyers, M. A., *3.4 Plastic Waves of Combined Stress*, Wiley, New York, 1994, pp. 77–80.

[15] Nicholas, T., "Tensile Testing of Materials at High Rates of Strain," *Experimental Mechanics*, Vol. 21, No. 5, May 1981, pp. 177–185.

[16] Kennan, Z., "Determination of the Constitutive Equations for 1080 Steel and Vascomax 300 AFIT/GAE/ENY/05-J05," Master's Thesis, Department of Aeronautics and Astronautics, Air Force Inst. of Technology, Wright-Patterson AFB, Dayton, OH, May 2005.

[17] Nicholas, T., Rajendran, A. M., and Grove, D. J., "Analytical Modeling of Precursor Decay in Strain-Rate Dependent Materials," *International Journal of Solids and Structures*, Vol. 23, No. 12, 1987, pp. 1601–1614.

[18] Nicholas, T., Rajendran, A. M., and Grove, D. J., "An Offset Yield Criterion for Precursor Decay Analysis," *Acta Mechanica*, Vol. 69, Nos. 1–4, Dec. 1987, pp. 205–218.

[19] Cinnamon, J. D., Palazotto, A. N., and Brar, N. S., "Further Refinement of Material Models for Hypervelocity Gouging Impacts," AIAA Paper 2006-2086, May 2006.

[20] Nguyen, M. C., Palazotto, A. N., and Cinnamon, J. D., "Analysis of Computational Methods for the Treatment of Material Interfaces," AIAA Paper 2005-2354, April 2005.

[21] Rickerd, G. S., Palazotto, A. N., and Cinnamon, J. D., "Investigation of a Simplified Hypervelocity Gouging Model," AIAA Paper 2005-2355, April 2005.

Chapter 6

Validation of Constitutive Models for Midrange Strain Rates

I. Introduction

TO VALIDATE the material flow models developed for VascoMax 300 and 1080 steel in Chapter 5, different approaches must be used for midrange strain rates (1/s to 10^3/s) and high strain rates (10^4/s to 10^6/s). This is because of the desire to ensure the constitutive model is valid across the entire range so that a CTH model of the hypervelocity gouging phenomenon is accurate. A single high-strain-rate experiment might mask the midrange contribution to the solution. A perfect example is the flyer plate experiments, in which a Johnson–Cook constitutive model was developed that matched those tests, but which abandoned the midrange split Hopkinson bar (SHB) data. This necessitated the use of the Zerilli–Armstrong model in order to fit both sets of experiments.

We know from the discussion in Chapter 3 that the flow stress model is critical to the solution accuracy for these hypervelocity impacts. While a portion of the gouge is at the high strain-rate range, a good portion of the deforming material is undergoing plasticity in the midrange strain-rate regime [1]. This will be illustrated in detail in Chapters 8 and 9.

Therefore, the midrange strain-rate range will be examined first. The constitutive models will be validated first with the SHB test model. A Taylor impact test was conducted to validate the models in a midrange strain-rate impact scenario. The validity of the flow models will be verified for this strain-rate regime.

II. Modeling the SHB Tests

To model the SHB tests presented in Chapter 5, a slightly different approach needs to be applied. Because of CTH being an Eulerian shock-wave code, creating model of the SHB test is difficult [2–6]. Materials in CTH are given properties, such as velocity, which apply to all material at the cell center. Therefore, establishing a nonmoving boundary condition at one end of a specimen is problematic. Consequently, a traditional finite element code with constitutive model modification capability was chosen for this particular part of the constitutive model validation.

An axisymmetric finite element model of the SHB tensile specimen was carried out using ABAQUS/Explicit® (version 6.5), which allows wave propagation to occur as a time function. This particular code is limited to the Johnson–Cook model. Therefore, an evaluation of the Johnson–Cook model developed in Table 5.3 from the SHB data exclusively is compared against the Johnson–Cook model appearing in Table 5.5, which incorporated the flyer plate data.

This numerical approach can thus be used to evaluate a strain pulse effect on the SHB test. The numerical analysis is performed for $\frac{1}{4}$ of the specimen because of the symmetry nature (see Fig. 6.1). The applied conditions for numerical analysis are the velocity of 5 m/s at the end of the specimen (equivalent to a strain rate of 10^3/s of the SHB test) and room temperature.

In using the models from Table 5.3, the posttest geometry of the SHB specimens was matched. In particular, the significant necking of the VascoMax 300 steel specimens was replicated [7–9] (see Fig. 6.2). Additionally, the relatively uniform plastic deformation of 1080 steel was also successfully recreated (see Fig. 6.3). Because necking occurs with VascoMax 300 and causes the simulation to terminate, the numerical results are associated with the results at 225 μs (from a total time span of 306 μs). The resulting deformation of SHB specimens matches the experimentally observed results of the SHB test well.

Figures 6.4 and 6.5 show the development of viscoplasticity and the associated temperature rise (shown in degrees Kelvin) in the deformed material. The localized temperature rise clearly appears at the center of VascoMax 300 tensile specimen. Furthermore, the amount of temperature increase from room temperature is around 800°C. On the other hand, the temperature rise for the 1080 steel specimen is relatively moderate (~160°C), and its distribution is fairly uniform.

This temperature concentration would make the specimen more prone to thermal softening in a localized region, leading to shear band creation. In the next section, a metallurgical confirmation of this will be presented.

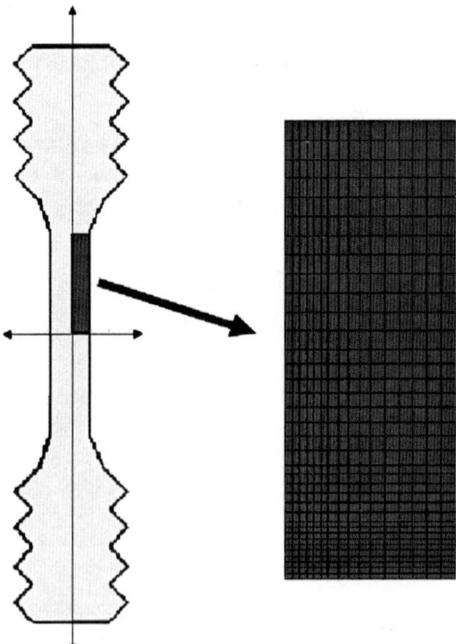

Fig. 6.1 Finite element model of SHB specimen.

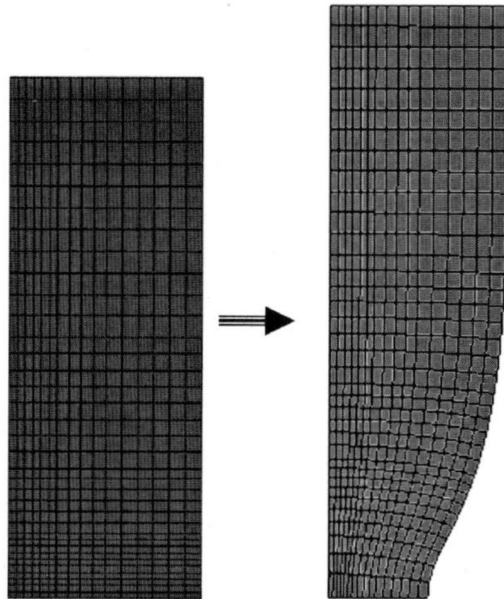

Fig. 6.2 Finite element results for VascoMax 300 SHB specimen.

When the Johnson–Cook model presented in Table 5.5 (which is the full-range strain-rate model, considering the flyer plate data) was applied to the same simulation, the results no longer matched the posttest measurements of the SHB specimens. The required linear relationship between effective flow stress and strain rate forced the model to miss these midrange strain-rate experimental points. Figure 6.6 highlights the difference in the results for VascoMax 300. In this figure, the effective stress is displayed in units of pascals. The flyer-plate modified

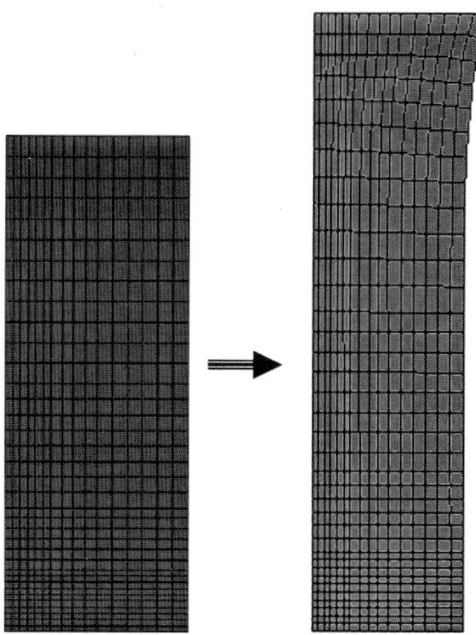

Fig. 6.3 Finite element results for 1080 steel SHB specimen.

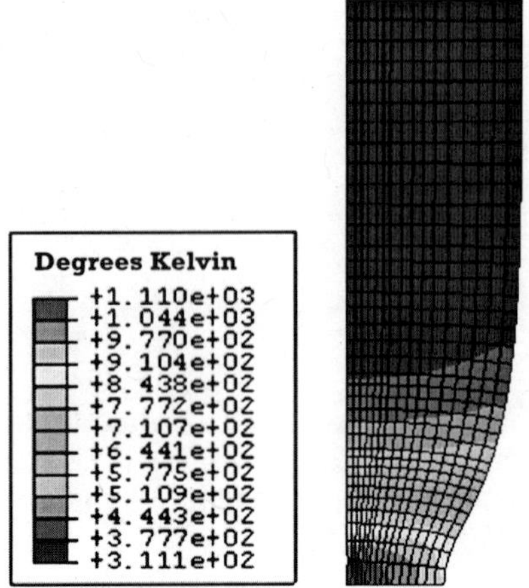

Fig. 6.4 Finite element thermal results for VascoMax 300 SHB specimen.

Johnson–Cook model appears on the right. Note that the necking is reduced because of the higher dynamic yield strength estimate.

When the constitutive model matches the SHB data better, the results are a better fit to the experimental measurements. Therefore, the more closely the full strain-rate range models are to those SHB data points, the better the midrange strain-rate results will be. Consequently, the Zerilli–Armstrong formulation is a superior choice in order to optimize the model's reflection of the entire range

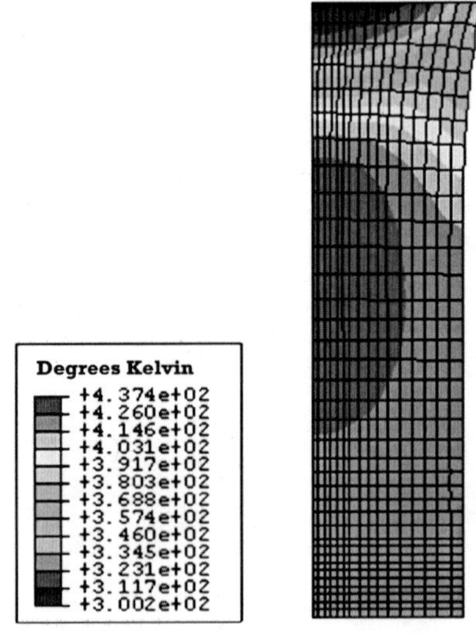

Fig. 6.5 Finite element thermal results for 1080 steel SHB specimen.

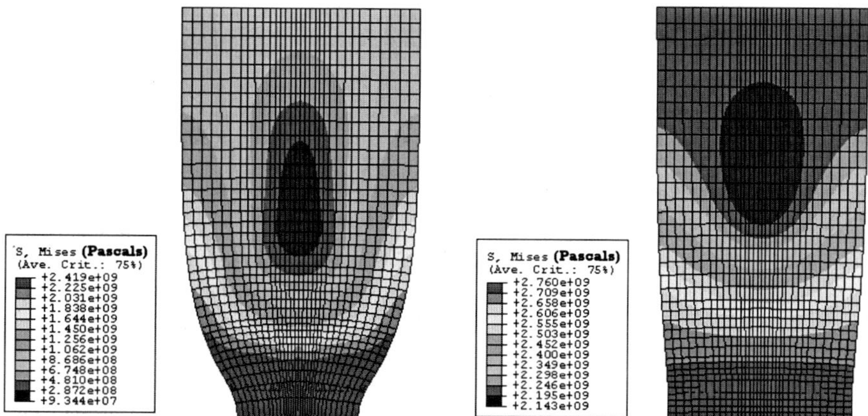

Fig. 6.6 Finite element result comparison for VascoMax 300 SHB specimen.

of experimental data. Because the full-range model, appearing in Table 5.5, was developed using the high strain-rate data, it reflects the entire strain-rate range. Additional experiments are presented in this work to justify the claim that the constitutive models developed are capable, within CTH, of successfully modeling high-energy impact events.

An additional result of this particular study was noting that the temperature concentration present in this finite element simulation of the SHB test provides an explanation for the unusual necking observed in the VascoMax 300 specimens. This result prompted a metallurgical study of the SHB specimens to verify this viscoplastic phenomenon.

III. Metallurgical Verification of SHB Model Results

To validate predictions regarding shear band formation within the plastically deformed SHB specimens, a metallurgical study was performed on the specimens, similar to that presented in Chapter 4 and [7, 10, 11]. Four specimens were sliced in half and examined using optical microscopy techniques. As noted in Chapter 4, the final polishing step of colloidal silica partially etched the surface of the specimens, making optical comparisons easier. Additionally, the same procedure was followed in examining 1080 steel specimens so that heat-affected zones could be identified if they exceeded the austenizing temperature.

Figure 6.7 shows a micrograph of a 1080 steel specimen (test T-12, 150°C, 500/s) at the fracture surface. There is a region of shear deformation, but there is no evidence of a heat zone and no localized region of shear damage. Note that the material damage is oriented at 45 deg to the axis of tension.

Figure 6.8 shows shear concentrations along the specimen length, away from the fracture surface. Again, the shear effects not being localized in a narrow band closely corresponds with the predictions made by the finite element results. Another 1080 specimen was examined (test T-38, 400°C, 1500/s) in which lower numbers of shear concentrations were observed.

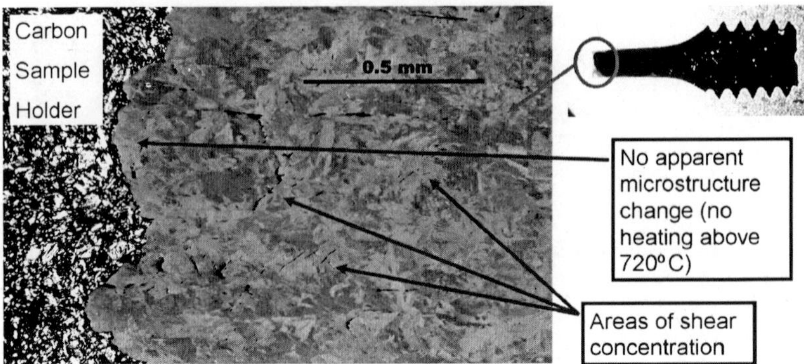

Fig. 6.7 SHB specimen T-12, as-polished, tip.

A somewhat different result was noted when an examination of the VascoMax 300 specimens was performed. Figure 6.9 shows a micrograph of the necked region (test T-4, room temperature, 500/s) of a VascoMax 300 SHB specimen. A large concentration of voids can be observed in the necked region and nowhere else. Furthermore, some shear concentrations were observed in Fig. 6.10 near the fracture surface, which did not occur elsewhere on the specimen.

These features contrast sharply with those away from the necked region, as shown in Fig. 6.11. Away from this region, there are no significant concentrations of voids and no shear concentrations. Another VascoMax 300 specimen (test T-27, 540°C, 1500/s), which represents a higher temperature and strain-rate, was examined, and the necked area of that specimen appears in Fig. 6.12. Note that the number of voids observed was less than test T-4. This again points to reduced shear localization as the initial temperature is increased.

These metallurgical results from the VascoMax 300 specimens compare favorably with the analysis presented in this work. Not only was the shear concentration localized in the necked region as predicted, but the unaffected nature of the

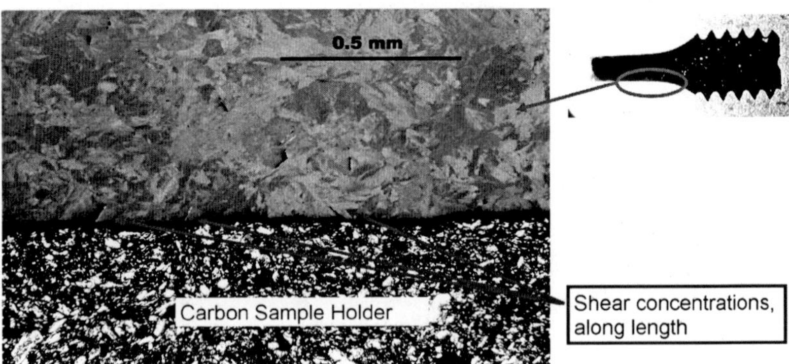

Fig. 6.8 SHB specimen T-12, as-polished, side.

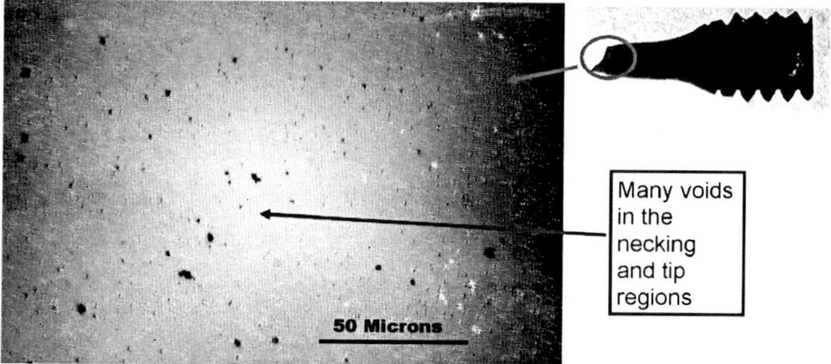

Fig. 6.9 SHB specimen T-4, as-polished, tip region.

microstructure away from that region was also reflected. Additionally, the higher initial temperature of test T-27 appears to have resulted in less shear generation within the resulting deformed material.

The micrographic examination of the SHB specimens appears to validate both our constitutive model and the resulting shear band prediction based on those constitutive relationships. Although this simulation could not be accomplished within CTH, it does provide verification that the constitutive model based on the SHB tests could replicate those same tests within a computational code. Again, the full-range constitutive model, developed in the Zerilli–Armstrong formulation, is as close as possible to those midrange strain-rate experimental data points while accurately estimating the high strain-rate regime.

The next set of experimental tests conducted was midrange strain-rate impact tests that could be modeled within CTH.

IV. Taylor Impact Tests

The Taylor impact test involves shooting a cylindrical projectile against a non-deforming target and making judgments concerning material characteristics based on the deformation of the projectile. Taylor first proposed this test in 1948 [12]

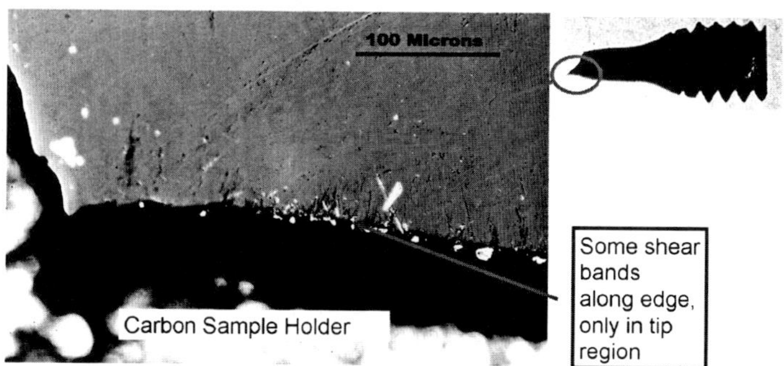

Fig. 6.10 SHB specimen T-4, as-polished, tip.

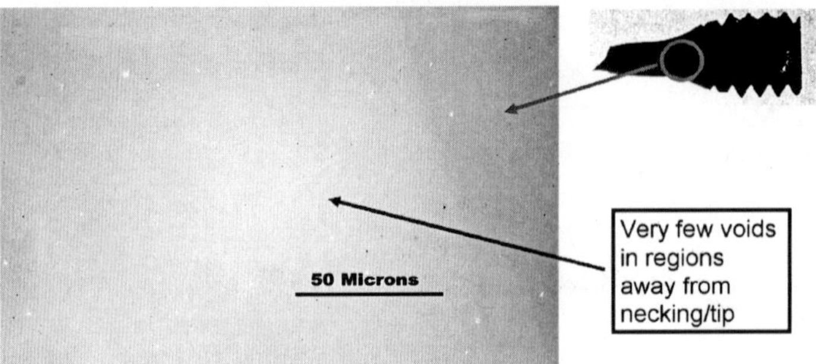

Fig. 6.11 SHB specimen T-4, as-polished, middle.

along with Whiffen [13]. Since then, the test has been used to estimate yield stress and to validate constitutive models in numerical codes (see Cinnamon et al. [14, 15], House [16], Wilson et al. [17], Jones et al. [18, 19] and Nicholas and Rajendran [20]). This is because the test is more available to investigators than other more costly experiments. Additionally, this test can generate impact strain rates on the order of 10^3/s, which serve to validate models created using the SHB test. Some researchers have used the Taylor test themselves to generate the material constitutive models [18, 19].

A. Taylor Test Overview

A Taylor impact test facility was created using the $\frac{1}{2}$-in. barrel light gas gun pictured in Fig. 6.13 (the back end of the gun, including the compressed gas bottle and firing solenoid) and 6.14 (the exit end of the barrel with a sabot stripper plate, the target, and fiducial). The target was a block of VascoMax 300 heat treated by the HHSTT to be metallurgically identical to the shoe material and then highly polished. A set of projectiles was manufactured with a nominal diameter of 6 mm and lengths of 30, 60, and 90 mm. They were constructed of 1080 steel (from as-received rail stock) and VascoMax 300 (heat treated in the same fashion as the

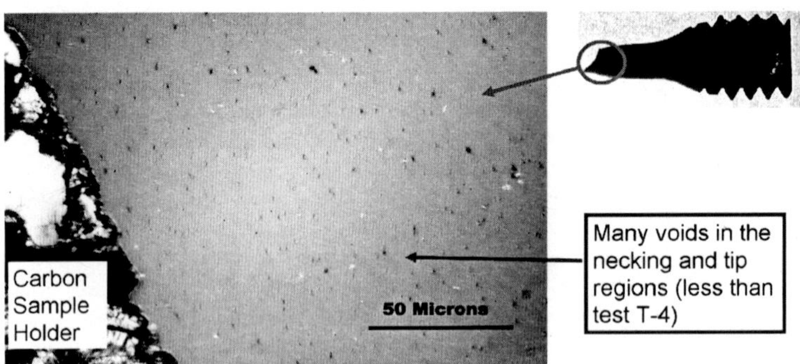

Fig. 6.12 SHB specimen T-27, as-polished, tip.

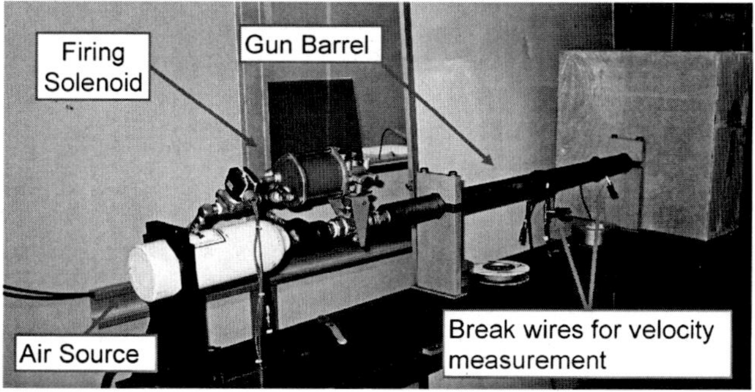

Fig. 6.13 Rear of light gas gun.

sled shoes). These materials, then, possess the same properties as the sled and rail materials in the HHSTT hypervelocity gouging problem.

Because of the chosen size of the cylinders, a sabot was necessary to hold the specimens in the $\frac{1}{2}$-in. barrel as they traveled down the gun. The sabots were fabricated from plastic and "stripped" off the projectile by the steel plate immediately past the barrel end. This plate allowed the projectile through and held the plastic sabot (although the sabots would typically break apart during the process after releasing the projectile). The projectiles would continue and impact the target. This impact was recorded via a high-speed digital camera with the capability to take a frame every 21 μs. This camera appears in Fig. 6.15.

The velocity of the projectile was measured with the use of trip wires across the barrel—two at various distances down the length of the barrel and one across the barrel exit. A computer would record the time difference between wire breaks,

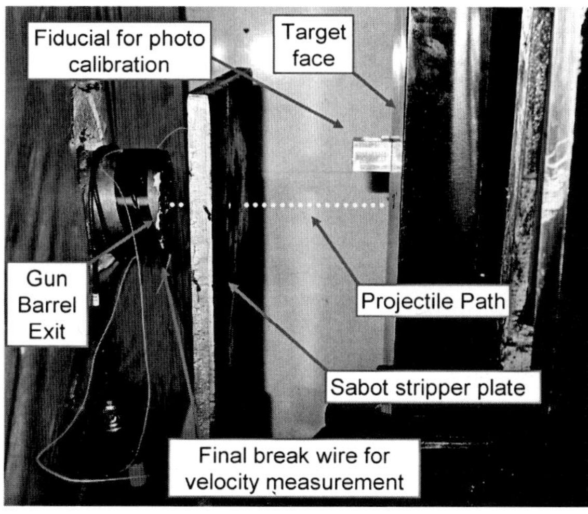

Fig. 6.14 Exit end of light gas gun.

Fig. 6.15 High-speed phantom digital camera and computer.

and a velocity could be computed. This velocity was used to validate our primary technique of reading the impact velocity using the high-speed photographs. The velocities from these two methods were within 10% of each other. The "fiducial" (of known dimensions) on the target face was used to calibrate distances in the digital photographs. Figure 6.16 illustrates an example high-speed photograph of a Taylor cylinder impact and the resulting deformation.

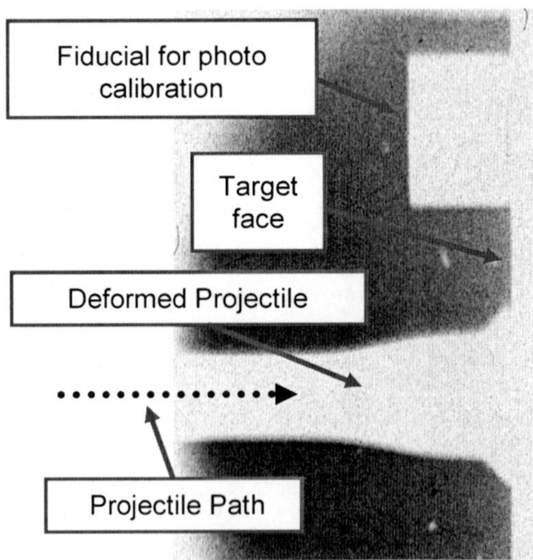

Fig. 6.16 Example high-speed photograph of Taylor impact.

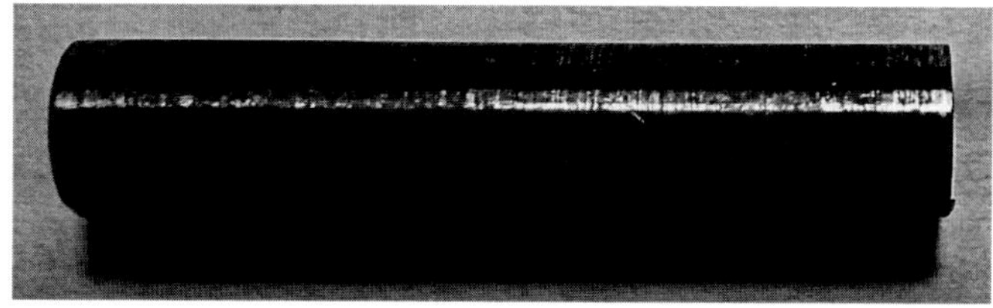

Fig. 6.17 Deformation of VascoMax 300 Taylor specimen (V10).

B. Taylor Test Results

Early in the process of conducting the Taylor tests, the VascoMax 300 exhibited a very low tolerance for an off-axis impact. The brittle nature of the material led to projectile fracturing at very low impact velocities (as low as 75 m/s). In testing VascoMax, a very close to normal impact and a low impact velocity were required. As the test procedure was improved and thereby the accuracy of the impact vector, more typical Taylor impacts were recorded. However, pushing the velocity up past 135 m/s resulted in fracture even with a normal impact. The 1080 steel projectiles experienced deformation in the typical fashion. Figures 6.17 and 6.18 show examples of posttest deformation.

Tables 6.1 and 6.2 summarize the Taylor impact results for VascoMax 300 and 1080 steel, where D_0 is the initial specimen diameter, L_0 is the initial specimen length, v_0 is the initial impact velocity (measured from the digital photographs), L_f is the specimen final length, D_f is the final mushroom diameter, and h_f is the undeformed section length (i.e., the remaining length of the cylinder that has not experienced any measurable diameter change). These quantities are illustrated in Fig. 6.19.

The high-speed camera provided an opportunity to examine the impact in a time-resolved manner. The capability of our equipment in this case, however, prevented a detailed investigation of that deformation. We simply did not possess sufficient resolution to track the plastic wave or measure mushroom diameter as a function of time after impact. The photographs offered a way to verify a normal impact and evaluate qualitative aspects of the event. The posttest measurements of the specimens were the only data in terms of deformation that we could reliably gather.

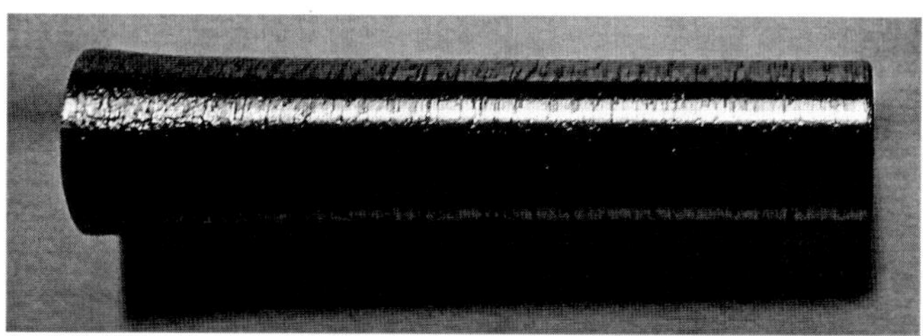

Fig. 6.18 Deformation of 1080 steel Taylor specimen (S6).

Table 6.1 Taylor impact results for VascoMax 300

Test shot	L_0, mm	D_0, mm	v_0, m/s	L_f, mm	D_f, mm	h_f, mm
V5	60.08	5.89	64	59.95	6.07	59.93
V6	60.06	5.93	76	59.78	6.1	58.9
V7	60.04	6.01	92	59.52	6.25	58.1
V9	30.02	6.08	83	29.81	6.22	27.5
V10	30	6.17	99	29.73	6.24	22.7
V11	59.91	5.88	107	59.4	6.01	53.4
V12	60.03	5.99	111	59.48	6.2	52.7
V13	89.92	5.93	111	89.04	6.12	83.3
V15	89.93	5.86	101	88.88	6.11	83.5

Selected test shots are presented in this work to illustrate the character of the impact event. These appear in Figs. 6.20–6.24. In some of these figures, the sabot material is seen behind the specimen because the sabot broke apart and came through the stripper plate. In others, the final trip wire is seen as it is being pushed out of the way by the cylinder. Each frame is separated by 21 μs (which allows us to adjust for the single shot in which the camera time stamp function malfunctioned). Figure 6.22 shows a normal impact of VascoMax 300 that resulted in a classic 45-deg fracture of the tip along the line of maximum shear stress.

C. Constitutive Model Validation via Taylor Tests

The Taylor impact tests were conducted prior to the flyer plate experiments. Therefore, two separate validations of the material flow models were performed. Initially, the Johnson–Cook material model appeared to be adequate for the effort to model hypervelocity impact, and so the material model in Table 5.3 was used.

Table 6.2 Taylor impact results for 1080 steel

Test shot	L_0, mm	D_0, mm	v_0, m/s	L_f, mm	D_f, mm	h_f, mm
S1	60	6	39	59.75	6	60
S2	60	6	134	57.46	6.6	41.32
S3	30	6	218	27.35	8.1	14.43
S5	29.96	6	207	27.44	7.9	16.44
S6	30	6	156	28.44	6.97	17.1
S7	30	6.01	263	25.94	9.59	11.29
S8	59.91	6	128	57.74	6.52	42.26
S9	59.95	5.89	148	57.19	6.6	39.96
S11	90.01	5.95	112	87.24	6.4	65.62

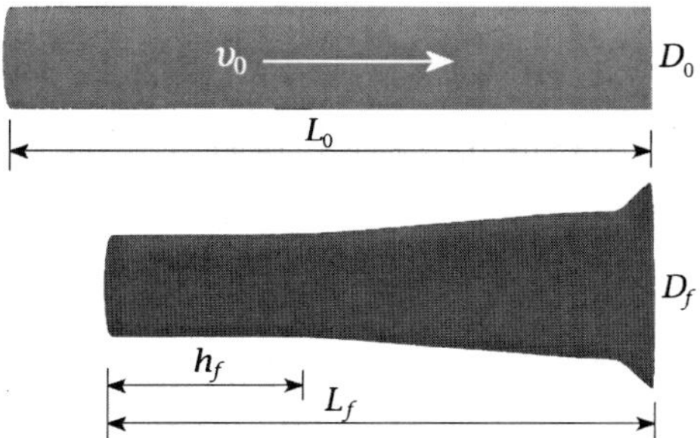

Fig. 6.19 Measured characteristics from a Taylor impact.

Fig. 6.20 High-speed camera photographs of specimen V10.

Fig. 6.21 High-speed camera photographs of specimen V15.

Fig. 6.22 High-speed camera photographs of specimen V14.

Fig. 6.23 High-speed camera photographs of specimen S6.

Fig. 6.24 High-speed camera photographs of specimen S8.

As an initial check on the validity of the Johnson–Cook coefficients determined by using the split Hopkinson bar data, a simple Lagrangian finite element Taylor impact test solver, authored by Cook (of the Johnson–Cook relationship) [21], was used. This solver allows the user to input an initial estimate of the Johnson–Cook parameters and the other material properties of a Taylor impact test and run the impact to see if the posttest geometry matches experimental data. Using the constants from the Hopkinson bar test, this solver showed excellent agreement between theoretical deformations and those seen in experimentation. This tool was used primarily because this code could complete a run in approximately 10 s on a desktop PC, whereas a similar impact in CTH requires approximately one hour using a state-of-the-art, multiple processor, parallel computing cluster.

Once we had established high confidence in these Johnson–Cook coefficients, the CTH model was constructed to perform the experimental tests. The CTH Taylor impact test model was created using a 0.002-cm mesh (as already discussed as mesh convergence value) in a two-dimensional axisymmetric implementation. The details of this can be found in Kennen [22] and [7, 8]. A more detailed discussion of the CTH modeling of these Taylor tests appears in Sec. VI. The results of this series of numerical simulations were as follows:

1) The CTH model achieves the correct D_f, L_f, and h_f within 2% of measured values and also matches the curvature of the mushroom for VascoMax 300.

2) The CTH model achieves the correct D_f, L_f, and h_f within 5% of measured values and also matches the curvature of the mushroom for 1080 steel.

3) The new 1080 steel material model is 100% more accurate, and the VascoMax 300 model is 50% more accurate than the previously used model [1, 23–27] for Iron and VascoMax 250, respectively.

After the flyer plate tests were completed and the full-range constitutive models were developed, these tests were revisited to ensure that the new Zerilli–Armstrong model would replicate these good results. The new model was slightly better in most cases and in no case worse than the previously reported match to the Taylor impact tests. The failure pressure used was the value arrived at using the method in Sec. III.C in Chapter 5.

Based on these results, the Zerilli–Armstrong material flow model presented in Table 5.5 was validated for the midrange strain-rate impact tests. As part of the investigation into the HHSTT gouging problem, material coatings (see Sec. III.F in Chapter 2) were also investigated using the Taylor test.

V. Study of HHSTT Coatings via Taylor Test

To study the two specific coatings currently in use at the HHSTT, experimental techniques to ascertain the friction coefficients at hypervelocity were explored. This effort proved problematic, in that hypervelocity friction studies do not exist. The highest friction values published between metals are in the neighborhood of 800 m/s. (See Bowden and Freitag [28], where they found at these velocities that the friction mechanism was caused by local adhesion and shearing between the contact surfaces.) Additionally, the test facility at the Air Force Research Laboratories also could not create a laboratory friction test above this velocity range. Therefore, a novel technique was created to coat Taylor impact specimens and use the test discussed in Sec. IV.A to compare the deformation of specimens [7, 8, 29].

A. Coated Taylor Impact Test Overview

A set of 1080 steel Taylor specimens was coated with iron oxide (hematite) and epoxy, which are the two coatings used on the rails at the HHSTT. An additional series of specimens was coated with an experimental material, known as "nanosteel" [30]. A series of Taylor impact tests were conducted against a Vasco-Max 300 target, with an effort made to examine deformation at various velocities and at those specific speeds that uncoated tests had been conducted. Figure 6.25 depicts the different Taylor impact specimens used in our testing. The iron oxide and epoxy coatings were applied by the HHSTT in the same manner as the rails

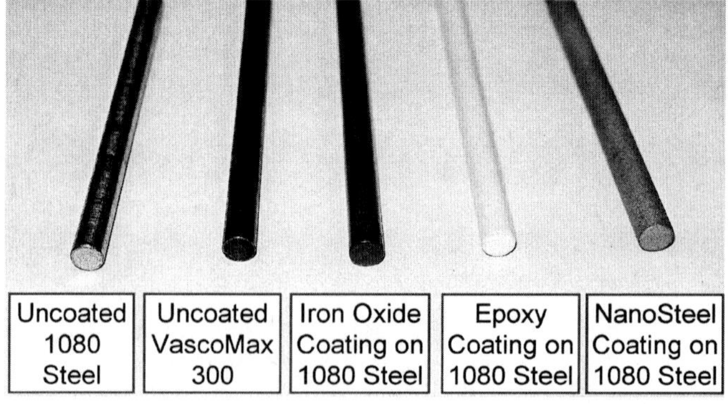

Fig. 6.25 Different Taylor impact specimens.

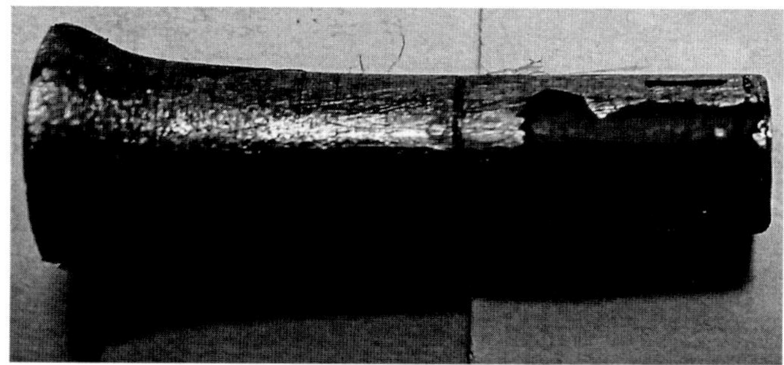

Fig. 6.26 Deformation of iron-oxide-coated 1080 steel Taylor specimen (I4).

are coated for sled test runs, specifically using the same techniques and thicknesses. The nanosteel specimens were coated by a NanoSteel, Inc., subcontractor (Engelhard) to the same thickness specifications.

B. Coated Taylor Impact Test Results

In the same manner as described in Sec. IV.A, a series of tests were conducted. The deformation was similar to those experienced for uncoated 1080 steel. Typical deformations can be seen in Figs. 6.26 and 6.27. Note that the coatings tended to fracture off of the sides of the specimens during the impact process.

The coated specimens exhibited a greater radial deformation as compared to the uncoated specimens. Tables 6.3–6.5 summarize the Taylor impact results for 1080 steel specimens coated with iron oxide, epoxy, and nanosteel.

Similar to Sec. IV.B, a series of selected test shots are presented in Figs. 6.28–6.32 to illustrate the impact event with coatings. Note in the photographs that the

Fig. 6.27 Deformation of epoxy-coated 1080 steel Taylor specimen (E5).

Table 6.3 Taylor impact results for iron oxide coated 1080 steel

Test shot	L_0, mm	D_0, mm	v_0, m/s	L_f, mm	D_f, mm	h_f, mm
I1	30.32	5.98	161	28.61	7.18	15.85
I2	60.19	5.92	130	57.9	6.65	37.4
I3	90.21	5.93	110	87.38	6.44	54.4
I4	30.28	5.93	243	26.25	9.7	10.7
I5	60.25	5.85	144	57.25	6.74	32.25

Table 6.4 Taylor impact results for epoxy-coated 1080 steel

Test shot	L_0, mm	D_0, mm	v_0, m/s	L_f, mm	D_f, mm	h_f, mm
E1	30.53	6.04	151	28.7	7.13	20.8
E2	60.55	5.96	128	57.97	6.65	47
E3	90.55	5.96	108	87.68	6.41	78.85
E4	30.51	5.97	243	26.31	10.2	14.25
E5	60.48	5.97	144	57.37	6.9	39.4

Table 6.5 Taylor impact results for nanosteel-coated 1080 steel

Test shot	L_0, mm	D_0, mm	v_0, m/s	L_f, mm	D_f, mm
N2	90.42	5.89	113	88.15	6.52
N4	60.28	5.95	118	58.76	6.44
N5	30.42	5.95	141	29.44	6.80
N6	30.47	5.95	254	26.85	9.50

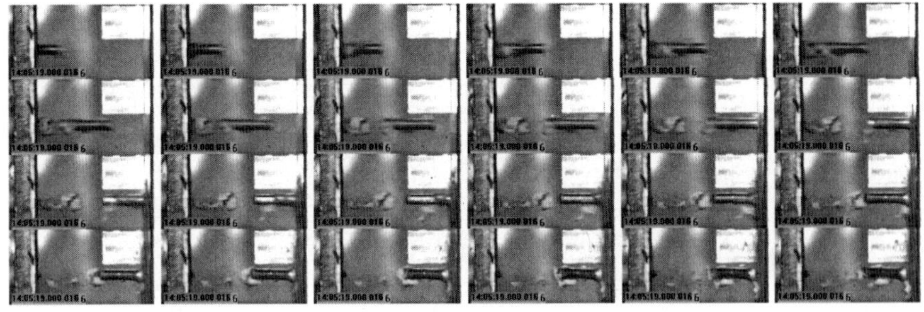

Fig. 6.28 High-speed camera photographs of specimen I4.

Fig. 6.29 High-speed camera photographs of specimen I5.

Fig. 6.30 High-speed camera photographs of specimen E4.

Fig. 6.31 High-speed camera photographs of specimen E5.

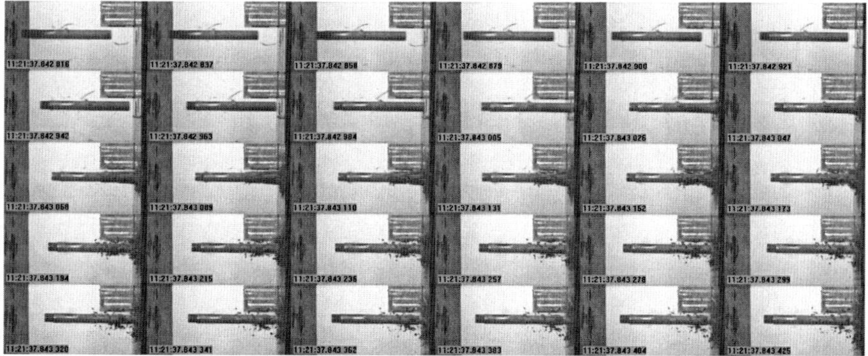

Fig. 6.32 High-speed camera photographs of specimen N4.

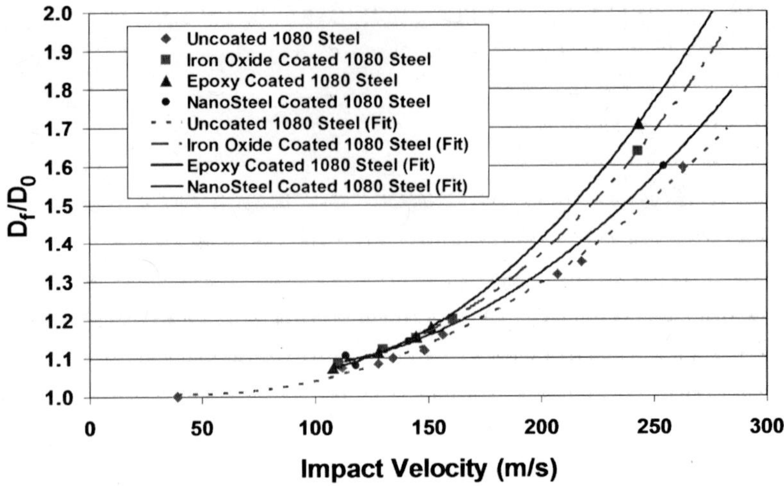

Fig. 6.33 Comparison of deformation for coated Taylor specimens.

coating has very little resistance to the elastic deformation wave and therefore fractures off the side of the specimen.

The coated specimens' deformation was characterized by greater mushroom growth (radial deformation against the target face) for the coated specimens vs the uncoated ones presented earlier. We know from estimates of the frictional coefficients that iron oxide has a lower coefficient than uncoated contact and that epoxy has a lower one than iron oxide [1]. This is graphically illustrated by comparing the diameter growth ratio (D_f/D_0) vs impact velocity. This comparison appears in Fig. 6.33. This clearly shows a relationship between increasing impact speed and the resulting mushroom diameter. Because the coatings act to reduce the friction against the target face, it is apparent that the epoxy has the lowest coefficient of friction, followed by iron oxide, nanosteel, and finally the noncoated state.

A simple one-dimensional analysis is presented in the next section in an effort to quantify the difference in the coefficient of friction between these coatings.

C. One-Dimensional Theory for Coating Comparison Using the Taylor Impact Test

To derive a simple relationship to compare the coefficients of friction between the three Taylor impact cases (no coating, coated with iron oxide, and coated with epoxy), we examine the deforming specimen in Fig. 6.34. In this depiction, F_I is the force of impact, F_f is the force of friction, F_x is the force of deformation in the x direction, and N is the normal force.

As we can see in the figure, the point of interest (the outside edge of the mushroom) can be thought of as experiencing the four forces depicted. To simplify this analysis, let us consider that F_I is constant over the time interval $0 \leq t \leq \Delta t_y$,

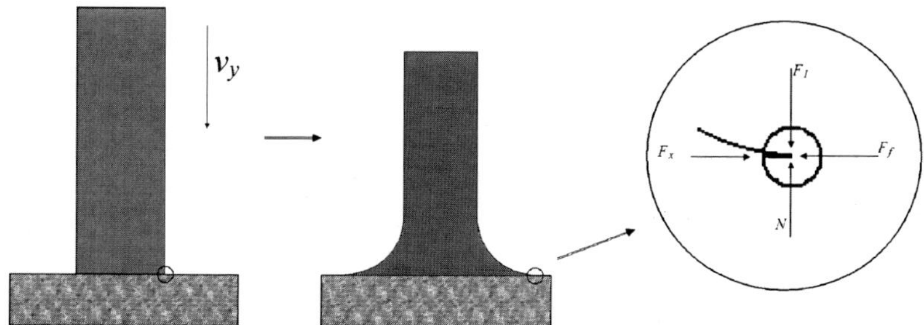

Fig. 6.34 Diagram of deforming Taylor specimen.

where t is time and Δt_y is the duration of the impact event in the vertical direction.

If we apply a simple impulse-momentum balance to the forces acting in the vertical direction, then it can be said that

$$F_I \Delta t_y = m \Delta v_y \tag{6.1}$$

where Δv_y is the impact velocity and m is the mass of the projectile. Because there is a force balance in the vertical direction, we know that

$$N = F_I = \frac{m \Delta v_y}{\Delta t_y} \tag{6.2}$$

If we consider that a state of dynamic equilibrium exists in the x direction, then $F_f = F_x$. Applying kinematics to the expression of F_x yields

$$F_x = m a_x = m \frac{\Delta v_x}{\Delta t_x} = m \frac{(\Delta s_x / \Delta t_x)}{\Delta t_x} = m \frac{\Delta s_x}{\Delta t_x^2} \tag{6.3}$$

where a_x is the acceleration of the deformation in the x direction, Δv_x is the change in deformation velocity in the x direction, Δt_x is the time duration of the deformation event in the x direction, and Δs_x is the deformation in the x direction. We also know that because $F_f = \mu N$ (where μ is the coefficient of friction), then

$$\mu = \frac{F_f}{N} = \frac{\Delta s_x \Delta t_y}{\Delta v_y \Delta t_x^2} \tag{6.4}$$

Making a computation for the values of μ becomes problematic in that we do not have accurate experimental values for Δt_x or Δt_y. We could analytically estimate Δt_y by using elastic wave speed theory and assert that the specimen remains in contact with the target for the length of time that it takes the elastic wave to travel to the specimen end and return as a tensile wave (and thereby pulling the specimen off the target). Making that claim, we arrive at

$$\Delta t_y = \frac{2L_0}{c_0} = \frac{2L_0}{\sqrt{E/\rho}} \tag{6.5}$$

We note, then, that Δt_y becomes a constant for a particular shot geometry if we are interested in making a comparison between the various coatings options on the specimens. Additionally, if we take the comparison at a fixed impact velocity, the Δv_y is also a constant. Therefore,

$$\mu \sim C \frac{\Delta s_x}{\Delta t_x^2} \tag{6.6}$$

where C is a constant. Now, let us assume that the deformation speed $\Delta s_x / \Delta t_x$ is a constant for a given geometry and impact velocity. That would lead to

$$\mu \sim \frac{C'}{\Delta t_x} \tag{6.7}$$

where C' is a constant. Note that this implies that a greater deformation Δs_x would require a proportionally greater Δt_x for the deformation speed to remain constant. A greater Δt_x would result in a lower relative value for μ, which makes intuitive sense. This constant deformation implies that

$$\Delta v_x = \frac{\Delta s_{x_1}}{\Delta t_{x_1}} = \frac{\Delta s_{x_2}}{\Delta t_{x_2}} \tag{6.8}$$

or

$$\Delta t_{x_2} = \frac{\Delta s_{x_2}}{\Delta s_{x_1}} \Delta t_{x_1} \tag{6.9}$$

Therefore, if we were to compare two shots of differing coatings,

$$\frac{\mu_1}{\mu_2} = \frac{C'/\Delta t_{x_1}}{C'/\Delta t_{x_2}} = \frac{\Delta t_{x_2}}{\Delta t_{x_1}} = \frac{(\Delta s_{x_2}/\Delta s_{x_1})\Delta t_{x_1}}{\Delta t_{x_1}} = \frac{\Delta s_{x_2}}{\Delta s_{x_1}} \tag{6.10}$$

Again, we note that if $\Delta s_{x_2} > \Delta s_{x_1}$, than this relationship would require that μ_1 be proportionally larger than μ_2. In a simple, one-dimensional sense then, we can compare the relative coefficients of friction between coating states by comparing the resulting mushroom diameters in the Taylor impact test.

By reexamining Fig. 6.33, we see that there is a distinct difference in mushroom diameter for the different Taylor specimens at the same impact velocity. The figure includes a simple polynomial fit to the experiment data. We can take the point at 243 m/s to compare the coefficients of friction, which corresponds to a velocity at which we have a couple of data points (and can extrapolate the other one). Additionally, at velocities higher that this range significant radial fractures in the mushroom occur. At this velocity (243 m/s)

$$\mu_{\text{epoxy}} = \frac{2.90\,\text{mm}}{4.23\,\text{mm}} \mu_{\text{uncoated}} = 0.69 \mu_{\text{uncoated}} \tag{6.11}$$

and

$$\mu_{\text{ironoxide}} = \frac{2.90\,\text{mm}}{3.77\,\text{mm}} \mu_{\text{uncoated}} = 0.77 \mu_{\text{uncoated}} \tag{6.12}$$

and

$$\mu_{\text{nanosteel}} = \frac{2.90 \text{ mm}}{3.15 \text{ mm}} \mu_{\text{uncoated}} = 0.92\mu_{\text{uncoated}} \tag{6.13}$$

Stated another way, with this simple analysis, the nanosteel coating appears to reduce the frictional effects by approximately 8%. The iron-oxide coating reduces that friction by another 15%. The epoxy coating reduces that friction by another 11%, for a total reduction of friction over the uncoated rail of 31%.

As we have seen in the literature, frictional effects play an important role in this hypervelocity gouging problem. This experimental work and analysis indicate that there is a significant reduction in friction using the coatings.

Although we have validated our constitutive models with respect to the uncoated Taylor tests, we need to also validate them in impacts involving the coating.

D. Constitutive Model Validation for Taylor Test Coated Specimens

Following the same procedure outlined in Sec. IV.C, the experimental tests were used to validate CTH models of the Taylor impact specimens impacting the VascoMax 300 target. In these cases, the model was modified to add a layer of coating (at a nominal thickness of 0.02 cm, which matches both the experimental specimens and the coating thickness used on the rail for the sled test at the HHSTT). A more detailed discussion of the CTH modeling of these Taylor tests appears in Sec. VI.

In the case of the coated specimens, a similar double set of validations for the flow models was performed—one validation using the SHB Johnson–Cook models [7, 8, 22, 29] and a final one using the Zerilli–Armstrong full-range model. CTH had experimentally based constitutive models and EOS models for both epoxy and iron oxide. Therefore, these CTH simulations of coated Taylor tests were also needed to validate the coatings models within the code.

CTH runs were very similar for these two validation cases. Remarkably good agreement was achieved between experimental results and the numerical predictions. The CTH model achieves the correct D_f, L_f, and h_f within 5% of measured values and also matches the curvature of the mushroom for 1080 steel with both iron-oxide and epoxy coatings.

Figure 6.35 provides an illustrative example in which CTH has replicated the final deformation of test E2 and the behavior of the coating, specifically the fracture of it off of the nose and sides of the projectile. The figure shows the posttest condition of test E2 and the CTH simulation of the impact (the projectiles traveled right to left in this particular depiction). Both the deformation of the projectile and the damage to the coating were accurately predicted by the simulation.

These results establish confidence in both the developed constitutive models for the 1080 and VascoMax 300 steels and the ability of CTH to model deformations with and without coatings.

VI. Modeling of Taylor Impact Tests in CTH

The validation of the material constitutive models within CTH necessitated a related study into how CTH handles the contact schemes before the Taylor impact model could be used.

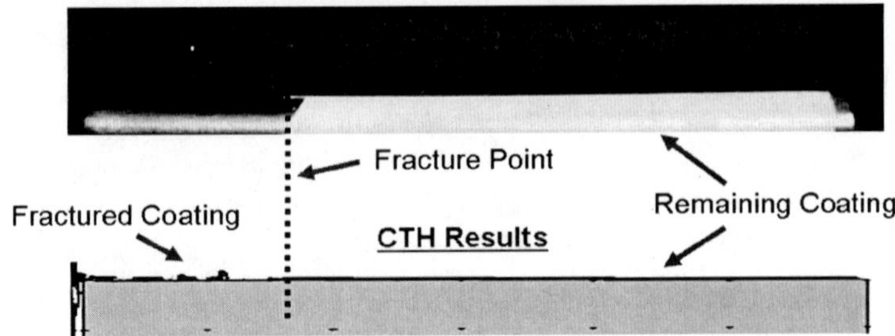

Fig. 6.35 Comparison of CTH results to projectile, test E2.

A. CTH Contact Schemes

One of the primary areas of study within the hypervelocity gouging phenomenon is the topic of friction, the heat generated by it, and how this interaction can be accurately reflected in a numerical model. CTH has several contact algorithms for describing material interactions on the interface that were carefully examined [31, 32].

There are three methods for defining the interface between materials in CTH. The first is the "no-slide" (default) condition. This approach assumes that the materials are joined upon contact, and the mixed cells (cells with two different materials within it) in the Eulerian mesh have strength characteristics weighted to the material volume fraction. This condition requires the materials to fail in shear for a sliding-type action to occur. To attempt to solve this difficulty, a second algorithm was developed called the "slide line." The slide line artificially sets the material shear strength of the mixed cells to zero, which allows sliding action to occur. This fluid-like behavior leads to undesirable results in a penetration-type impact in that the typically harder projectile experiences erosion during penetration. To correct this, a third algorithm was developed, known as the "boundary-layer" approach [6]. This algorithm moves the slide line into the target material in order to preserve the integrity of the penetrator. This approach is the only one in which the user can explicitly set a coefficient of friction between the materials.

To study these algorithms, a simple sliding model was developed in which a rod of VascoMax 250 slides within a stationary cylinder of iron (these being the closest materials in CTH to VascoMax 300 and 1080 steel at the time of the study) under continuous contact. The interior rod was not given a velocity vector to allow collision and should not have interacted significantly with the target.

The slide-line algorithm developed significant numerical instability and created shear stresses within the target far away from the material interface. The result of the simulations was nonphysical stresses and thermodynamic characteristics. The no-slide and boundary-layer algorithms produced similar results, with some numerical noise, but much better than the slide-line approach.

A further investigation was conducted in which a normal penetrating impact was evaluated using these various contact schemes. The outcome of this study was to determine that the slide line created unrealistic results and that careful use of both the no-slide and boundary-layer algorithms could yield good results. Although the slide-line approach might be valid in some kinds of problems, the hypervelocity gouging problem requires a judicious application of the interface conditions.

The boundary-layer algorithm, however, was found to be valid only in a two-dimensional axisymmetric case and only using single-processor CTH computations. This limitation makes that particular algorithm of little use in the simulation of a full shoe/rail geometry model, which requires a two-dimensional plane-strain, multiprocessor mode. This particular limitation was unknown to previous investigators, with unknown impact on the results derived from using the boundary-layer algorithm in this multiprocessor, two-dimensional plane-strain mode.

Therefore, the default contact scheme is the best choice to model hypervelocity impact. Eulerian hydrocodes, in general, have difficulty modeling sliding interfaces, friction, and contact [33]. However, the no-slide scheme selected offers the best opportunity to generate good results using CTH.

B. CTH Taylor Test Model

Having established the optimum contact algorithm for the simulation of the Taylor impact test, a model was created with CTH. As already mentioned, a mesh convergence study was conducted by Szmerekovsky, in which the 0.002-cm cell size was found to be the limit of continuum mechanics for these steels and where the solution converged [1].

Figure 6.36 depicts the mesh used to simulate the Taylor specimens. The boundary conditions on the edges of the mesh were selected to be hydrodynamic conditions, which allow stress waves to pass (emulating a semi-infinite edge).

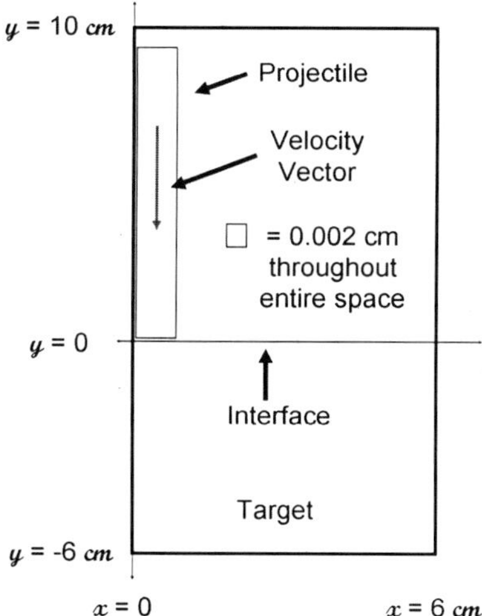

Fig. 6.36 CTH mesh for Taylor impact test simulation.

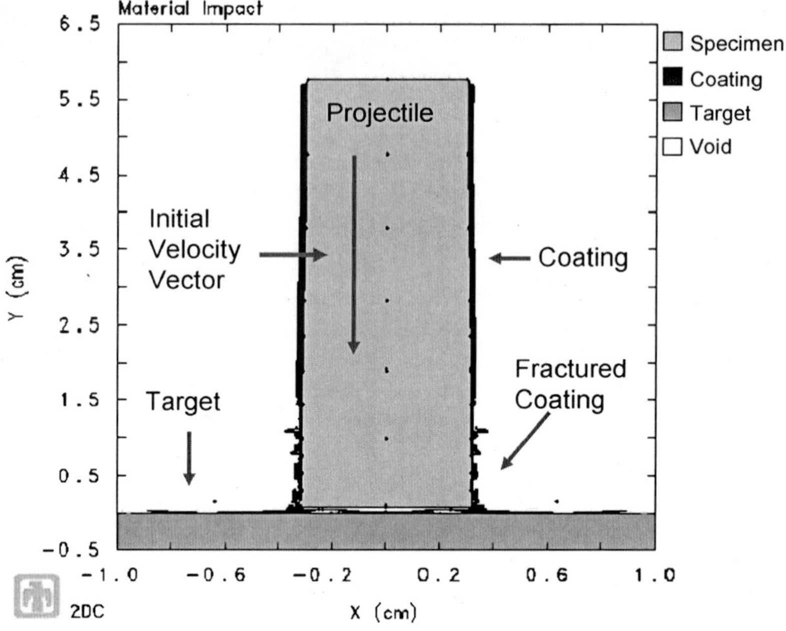

Fig. 6.37 Example CTH solution for Taylor impact test.

Lagrangian tracer points were included along the specimen edges to track material flow and to measure final diameter, length, and undeformed section length. A typical CTH simulation appears in Fig. 6.37. In this particular case, the specimen has impacted the target and bounced back away. The fragments in the field are from the coating fracturing off of the specimen. An important note here is that the presence of the coating aided the mushroom development in the simulation in the same manner that it did in the experiment, acting as a sacrificial shear layer that reduced the effective friction between the projectile and the target.

Figure 6.38 illustrates one of the many possible plots available from CTH. In this case, the midrange strain rates of 10^4/s and below are verified by observing the strain rate of the early stages of deformation. After this initial stage, the strain rates drop immediately to the 10^2/s to 10^3/s range and then decrease as the event continues.

C. CTH Modeling Conclusion

Based on a study conducted to ascertain the best contact scheme for use in simulation impact scenarios, the default no-slide condition was shown to be the most suitable. Additionally, previous work with the boundary-layer algorithm might, in fact, be invalid because of the implementation of the algorithm within the CTH code. Finally, a CTH Taylor test model was developed and was successfully used to simulate the impact events.

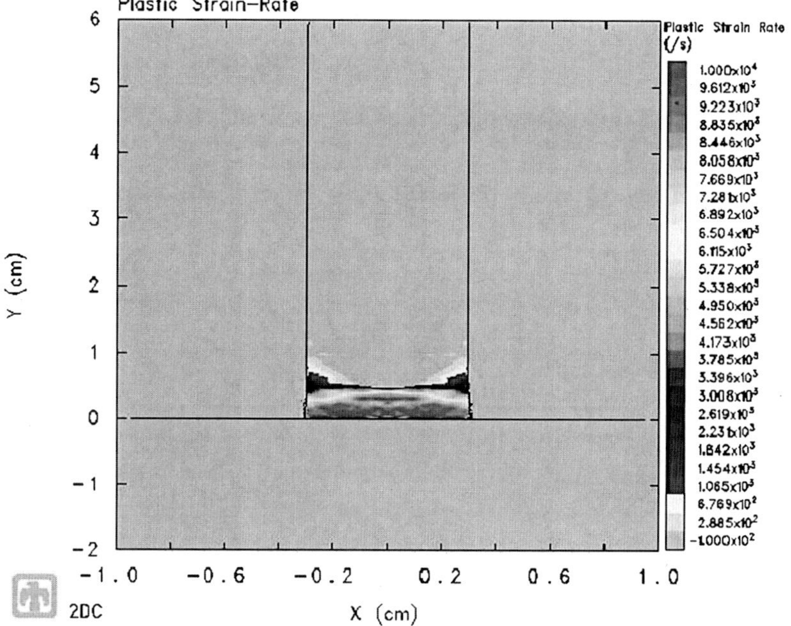

Fig. 6.38 Example CTH strain-rate solution for Taylor impact test.

VII. Summary

The midrange strain-rate regime of the material constitutive models developed from experimentation in Chapter 5 were validated in a simulation of the SHB tests using a Lagrangian finite element code. The thermal characteristics that developed from these flow models, with respect to the creation of shear bands, were verified metallurgically. The flow models therefore demonstrated good fidelity in predicting material behavior in that type of test.

To transition the validation effort into the impact testing realm, a series of Taylor impact tests were conducted (with and without coatings). These tests were successfully simulated using CTH, which validated not only the midrange portion of the full constitutive models, but also the CTH models for the coatings. These CTH models were developed after a tangential study concluded that the no-slide contact scheme in CTH was the most accurate for the simulation of impacts.

Whereas the constitutive models have been validated at the midrange strain-rate values, the hypervelocity gouging impact at the HHSTT also includes the high strain-rate regime. Therefore, to fully validate these models for application to that problem, a laboratory hypervelocity impact test was developed.

References

[1] Szmerekovsky, A. G., "The Physical Understanding of the Use of Coatings to Mitigate Hypervelocity Gouging Considering Real Test Sled Dimensions AFIT/DS/ENY

04-06," Ph.D. Dissertation, Department of Aeronautics and Astronautics, Air Force Inst. of Technology, Wright-Patterson AFB, Dayton, OH, Sept. 2004.

[2] Hertel, E. S., Bell, R. L., Elrick, M. G., Farnsworth, A. V., Kerley, G. I., McGlaun, J. M., Petney, S. V., Silling, S. A., Taylor, P. A., and Yarrington, L., "CTH: A Software Family for Multidimensional Shock Physics Analysis," *Proceedings of the 19th International Symposium on Shock Waves*, Sandia National Laboratories Rept. SAND-92-2089C, pp. 377–382.

[3] van Leer, B., "Towards the Ultimate Conservative Difference Scheme IV. A New Approach to Numerical Convection," *Journal of Computational Physics*, Vol. 23, No. 2, Aug. 1997, p. 276.

[4] McGlaun, J. M., Thompson, S. L., and Elrick, M. G., "CTH: A Three-Dimensional Shock Wave Physics Code," *International Journal of Impact Engineering*, Vol. 10, Issues 1–4, 1990, pp. 351–360.

[5] Silling, S. A., "Stability and Accuracy of Differencing Schemes for Viscoplastic Models in Wavecodes," Sandia National Labs., Technical Rept. SAND 91-0141, Albuguergue, NM, March 1991.

[6] Silling, S. A., "An Algorithm for Eulerian Simulation of Penetration," *New Methods in Transient Analysis*, PVP-Vol. 246, AMD-Vol. 143, ASME, Anaheim, CA, June 1992, pp. 123–128.

[7] Cinnamon, J. D., and Palazotto, A. N., "Refinement of a Hypervelocity Model for the Rocket Sled Test," *Proceedings of the 2005 ASME International Mechanical Engineering Congress and Exposition*, ASME Paper IMECE 2005-80004, ASME, New York, Nov. 2005.

[8] Cinnamon, J. D., Palazotto, A. N., and Kennan, Z., "Material Characterization and Development of a Constitutive Relationship for Hypervelocity Impact of 1080 Steel and VascoMax 300," *International Journal of Impact Engineering*, Vol. 33, Nos. 1–12, 2006, pp. 180–189.

[9] Cinnamon, J. D., Palazotto, A. N., Kennan, Z., Brar, N. S., and Bajaj, D., "Johnson-Cook Strength Model Constants for VascoMax 300 and 1080 Steels," *Proceedings of the 14th APS Topical Conference on Shock Compression of Condensed Matter*, Vol. 845, American Physical Society, College Park, MD, 2006, pp. 709–712.

[10] Cinnamon, J. D., and Palazotto, A. N., "Metallographic Examination of Thermal Effects in Hypervelocity Gouging," *Proceedings of the 2005 ASME Pressure Vessels and Piping Division Conference*, ASME Paper PVP 2005-71613, July 2005.

[11] Cinnamon, J. D., and Palazotto, A. N., "Metallographic Examination and Validation of Thermal Effects in Hypervelocity Gouging," *ASME Journal of Pressure Vessel Technology*, Vol. 129, No. 1, 2007, pp. 133–141.

[12] Taylor, G. I., "The Use of Flat-Ended Projectiles for Determining Dynamic Yield Stress, I. Theoretical Considerations," *Proceedings of the Royal Society of London, Series A*, Vol. 194, No. 1038, Sept. 1948, p. 289.

[13] Wiffen, A. C., "The Use of Flat-Ended Projectiles for Determining Dynamic Yield Stress, II. Tests on Various Metallic Materials," *Proceedings of the Royal Society of London, Series A*, Vol. 194, No. 1038, Sept. 1948, p. 300.

[14] Cinnamon, J. D., Jones, S. E., Foster, J. C., and Gillis, P. P., "An Analysis of Early Time Deformation Rate and Stress in the Taylor Impact Test," *Proceedings of the Sixth International Conference on the Mechanical Behavior of Materials*, Vol. 1, ASME, New York, 1991, pp. 337–342.

[15] Cinnamon, J. D., Jones, S. E., House, J. W., and Rule, W. K., "Validating the High Strain-Rate Strength Estimates Generated from High-Speed Film Data and a Revised Elementary Theory for the Taylor Impact Test," *Proceedings of the ASME Pressure Vessel and Piping Conference*, Vol. 1, ASME, New York, 2000, pp. 343–348.
[16] House, J. W., "Taylor Impact Testing," Air Force Armament Lab., Technical Rept. AFATL-TR-89-41, Eglin AFB, FL, Sept. 1989.
[17] Wilson, L. L., House, J. W., and Nixon, M. E., "Time Resolvable Deformation from the Cylinder Impact Test," Air Force Armament Lab., Technical Rept. AFATL-TR-89-76, Eglin AFB, FL, Nov. 1989.
[18] Jones, S. E., Drinkard, J. A., Rule, W. K., and Wilson, L. L., "An Elementary Theory for the Taylor Impact Test," *International Journal of Impact Engineering*, Vol. 21, Issues 1–2, Jan.–Feb. 1998, pp. 1–13.
[19] Jones, S. E., Maudlin, P. J., and Foster, J. C., "Constitutive Modeling Using the Taylor Impact Test," *Symposium on High Strain Rate Effects on Polymer, Metal and Ceramic Matrix Composites and Other Advanced Materials*, ASME Congress and Exposition, Vol. 48, San Francisco, CA, Nov. 1995, pp. 161–166.
[20] Nicholas, T., and Rajendran, A. M., "Material Characterization at High Strain Rates," *High Velocity Impact Dynamics*, Wiley, New York, 1990, pp. 127–296.
[21] Cook, W. H., "2D Axisymmetric Lagrangian Solver for Taylor Impact with Johnson-Cook Constitutive Model," Air Force Research Lab., Technical Rept. AFRL-MN-EG-TR-2000-7026, Eglin AFB, FL, April 2000.
[22] Kennen, Z., "Determination of the Constitutive Equations for 1080 Steel and VascoMax 300 AFIT/GAE/ENY/05-J05," Master's Thesis, Department of Aeronautics and Astronautics, Air Force Inst. of Technology, Wright-Patterson AFB, Dayton, OH, May 2005.
[23] Szmerekovsky, A. G., and Palazotto, A. N., "Structural Dynamics Considerations for a Hydrocode Analysis of Hypervelocity Test Sled Impacts," *AIAA Journal*, Vol. 44, No. 6, June 2006, pp. 1350–1359.
[24] Szmerekovsky, A. G., Palazotto, A. N., and Baker, W. P., "Scaling Numerical Models for Hypervelocity Test Sled Slipper-Rail Impacts," *International Journal of Impact Engineering*, Vol. 32, No. 6, 2006, pp. 928–946.
[25] Szmerekovsky, A. G., Palazotto, A. N., and Ernst, M. R., "Numerical Analysis for a Study of the Mitigation of Hypervelocity Gouging," AIAA Paper 2004-1922, March 2004.
[26] Szmerekovsky, A. G., Palazotto, A. N., and Ernst, M. R., "Numerical Analysis for a Study of the Mitigation of Hypervelocity Gouging," *Dayton-Cincinnati Aerospace Science Symposium*; modified and published as AIAA Paper 2004-1922, March 2004.
[27] Szmerekovsky, A. G., Palazotto, A. N., and Cinnamon, J. D., "An Improved Study of Temperature Changes During Hypervelocity Sliding High Energy Impact," AIAA Paper 2006-2090, May 2006.
[28] Bowden, F. P., and Freitag, E. H., "The Friction of Solids at Very High Speeds," *Proceedings of the Royal Society of London, Series A*, Vol. 248, Issue 1254, 1958, pp. 350–367.
[29] Blomer, M. A., "An Investigation for an Optimum Hypervelocity Rail Coating AFIT/GAE/ENY/05-J01," Master's Thesis, Department of Aeronautics and Astronautics, Air Force Inst. of Technology, Wright-Patterson AFB, Dayton, OH, May 2005.
[30] Goldman, L., "Steely Dan," *Forbes Magazine*, 23 Dec. 2002, pp. 12–13.

[31] Nguyen, M. C., "Analysis of Computational Methods for the Treatment of Material Interfaces AFIT/GAE/ENY 05-M15," Master's Thesis, Department of Aeronautics and Astronautics, Air Force Inst. of Technology, Wright-Patterson AFB, Dayton, OH, March 2005.

[32] Nguyen, M. C., Palazotto, A. N., and Cinnamon, J. D., "Analysis of Computational Methods for the Treatment of Material Interfaces," AIAA Paper 2005-2354, April 2005.

[33] Anderson, C. F., "An Overview of the Theory of Hydrocodes," *International Journal of Impact Engineering*, Vol. 5, Issues 1–4, 1987, pp. 33–59.

Chapter 7

Scaled Laboratory Hypervelocity Gouging Test

I. Introduction

VALIDATING the developed Zerilli–Armstrong constitutive models for VascoMax 300 and 1080 steel within the CTH hydrocode is necessary in order to ensure that simulations of the HHSTT sled scenario are accurate. Unfortunately, precise conditions at the point of gouging in the field are not known. That is, the HHSTT facility does not have the instrumentation arranged so that, where gouges occur, the sled parameters are recorded. Additionally, intentional gouging is not possible to arrange for analytical purposes. Therefore, the development of a laboratory hypervelocity gouging test was undertaken.

The dimensionality of the HHSTT sled makes full-scale gouge tests prohibitive. To create a laboratory gouging test, the sled scenario was scaled down to in order to test in the laboratory. The purpose was to evaluate the HHSTT scenario in a scaled experiment. A mathematically rigorous scaling (via the Buckingham pi technique) led to a geometry that was beyond the range of available laboratory facilities. To adjust for this eventuality, a one-dimensional penetration theory was developed to ensure laboratory tests would create gouging.

With this background, a series of hypervelocity gouging tests were conducted for the purpose of creating cases for CTH to match with our new constitutive models. The goal was to validate CTH's ability to generate correct predictions for hypervelocity gouging impacts, prior to its use in modeling the HHSTT scenario.

II. Scaled Gouging Test Development

In developing a laboratory hypervelocity gouging scenario, in which to examine this phenomenon of gouging and to establish test parameters to simulate within CTH, a mathematical scaling approach was utilized. The well-known Buckingham pi technique, which has been applied on this type of problem previously [1–4], was adopted. The goal of this effort was to arrive at test parameters that could be replicated by the gun facility that was available. A very detailed presentation of this technique can be found in [4] and [5].

According to the Buckingham pi theorem, if a physical law consists of a number (m) of quantities $\{q_i\}$, where $i = 1 \ldots m$, that have dimension and are products and powers of j independent fundamental dimensions L_j, then a unit free fundamental law can be defined as

$$f(q_1, q_2, q_3, \ldots, q_m) = 0 \qquad (7.1)$$

where m is the number of dimensioned quantities to be used in the analysis [6, 7]. A fundamental dimension is a quantity that is used to describe a dimensioned quantity. There are many different fundamental systems that can be used such as force, length, time (FLT) and mass, length, time (MLT). Take pressure, for example, in the FLT system; pressure would be represented as FL^{-2}. In the MLT system, pressure is represented as $ML^{-1}T^{-2}$. It must be ensured that the fundamental dimensions alone can describe all dimensioned quantities.

As just mentioned, it is possible to represent any dimensioned quantity as a product of fundamental dimensions raised to some power:

$$q_i = \left[L_1^{d_1} L_2^{d_2} \cdots L_n^{d_n}\right]_i \tag{7.2}$$

where q_i is a dimensioned quantity, L_j is a fundamental dimension, and d_k is the power the fundamental dimension is raised to. The dimensioned quantities can then be combined to form invariant pi quantities:

$$\Pi = (q_1)^{\alpha_1} (q_2)^{\alpha_2} \cdots (q_m)^{\alpha_m} \tag{7.3}$$

where the α_i is an exponent to be determined. It then follows that

$$\Pi = \left(L_1^{d_1} L_2^{d_2} \cdots L_n^{d_n}\right)_1^{\alpha_1} \left(L_1^{d_1} L_2^{d_2} \cdots L_n^{d_n}\right)_2^{\alpha_2} \cdots \left(L_1^{d_1} L_2^{d_2} \cdots L_n^{d_n}\right)_m^{\alpha_m} \tag{7.4}$$

Rearranging this equation so that all of the L_i quantities are together leads to

$$\Pi = (L_1)^{\beta_1} (L_2)^{\beta_2} \cdots (L_n)^{\beta_n} \tag{7.5}$$

where the exponents β can be described as

$$\begin{Bmatrix} \beta_1 \\ \beta_2 \\ \vdots \\ \beta_n \end{Bmatrix} = \begin{bmatrix} d_{11} & d_{12} & \cdots & d_{1m} \\ d_{21} & d_{22} & & d_{2m} \\ \vdots & & \ddots & \vdots \\ d_{n1} & d_{n2} & \dots & d_{nm} \end{bmatrix} \begin{Bmatrix} \alpha_1 \\ \alpha_2 \\ \vdots \\ \alpha_m \end{Bmatrix} \tag{7.6}$$

Mathematically, $\{\alpha\}$ must exist in the null space of the dimension matrix $[D]$ for the physical law to be dimensionally consistent. This requires that $\{\beta\} = \{0\}$. This requirement forces the solution of Eq. (7.6) to give the products of dimensioned quantities that must remain invariant between models [7].

Also according to the theorem, if there are m dimensioned quantities and r fundamental dimensions, then there are $k = m - r$ independent dimensionless quantities. In the MLT system there will be $r = 3$ fundamental dimensions [7].

Careful selection of the variables to be used within the Buckingham pi approach is required. Those characteristics, such as material density, which can be expressed as functions of other chosen parameters, are removed from consideration. Because the authors wish to scale the shoe/rail geometry, but still experiment with the materials at the HHSTT (VascoMax 300 and 1080 steel), some of the material properties (such as the wave speed of the material) are removed from consideration also. If that was not done, the Buckingham pi theorem would dictate an experimental test in which two different materials (which result from the scaling of the material properties) should be shot in our laboratory hypervelocity scenario. Therefore, those properties that cannot be scaled are removed from consideration, and the dimension scaling rule is made mathematically more sound (Barenblatt, G. I., private communication concerning scaling the HHSTT problem, 2005 ASME International Congress, Orlando, FL, Nov. 2005).

Previous modeling of the sled/rail interaction was done in a plane-strain manner [4, 8–12]. This choice and the implications to our scenario will be discussed later in this work in detail. At this point, however, the dimension of width (into the depth of a plane-strain implementation) is removed from consideration. Additionally, the rail dimensions are not scaled because the rail appears as an infinite half-plane of material to the shoe (or scaled impact projectile) over the time scale of a gouging impact (on the order of 10–20 μs). Therefore, taking the minimum number of fundamental characteristics from the sled/rail geometry, we arrive at the selected dimensioned quantities appearing in Table 7.1.

With these choices, the invariant parameter Π then becomes

$$\Pi = (m)^{\alpha_1} (l)^{\alpha_2} (d)^{\alpha_3} (u_x)^{\alpha_4} \left(u_y\right)^{\alpha_5} \left(\sigma_{y,c}\right)^{\alpha_6} (E_m)^{\alpha_7} (G_o)^{\alpha_8} (t)^{\alpha_9} \tag{7.7}$$

In fundamental dimension form, Eq. (7.4) becomes

$$\Pi = (M)^{\alpha_1} (L)^{\alpha_2} (L)^{\alpha_3} \left(LT^{-1}\right)^{\alpha_4} \left(LT^{-1}\right)^{\alpha_5} \times \left(ML^{-1}T^{-2}\right)^{\alpha_6} \left(ML^{-1}T^{-2}\right)^{\alpha_7} \left(ML^{-1}T^{-2}\right)^{\alpha_8} (T)^{\alpha_9} \tag{7.8}$$

This reduces to

$$\Pi = (M)^{\beta_1} (L)^{\beta_2} (T)^{\beta_3}$$

where

$$\begin{aligned} \beta_1 &= \alpha_1 + \alpha_6 + \alpha_7 + \alpha_8 \\ \beta_2 &= \alpha_2 + \alpha_3 + \alpha_4 + \alpha_5 - \alpha_6 - \alpha_7 - \alpha_8 \\ \beta_3 &= -\alpha_4 - \alpha_5 - 2\alpha_6 - 2\alpha_7 - 2\alpha_8 + \alpha_9 \end{aligned} \tag{7.9}$$

Table 7.1 Buckingham pi dimensioned quantities

Dimensioned quantity	Fundamental dimensions
Mass m	M
Height d	L
Length l	L
Horizontal velocity u_x	LT^{-1}
Vertical velocity u_y	LT^{-1}
Compressive yield strength $\sigma_{y,c}$	$ML^{-1}T^{-2}$
Elastic modulus E_0	$ML^{-1}T^{-2}$
Shear modulus G_0	$ML^{-1}T^{-2}$

Setting the values of β to zero and solving for $m = \alpha_1$, $l = \alpha_2$, and $u_x = \alpha_4$, one obtains

$$\begin{aligned}\alpha_1 &= -\alpha_6 - \alpha_7 - \alpha_8 \\ \alpha_2 &= -\alpha_3 + 3\alpha_6 + 3\alpha_7 + 3\alpha_8 - \alpha_9 \\ \alpha_4 &= -\alpha_5 - 2\alpha_6 - 2\alpha_7 - 2\alpha_8 + \alpha_9\end{aligned} \tag{7.10}$$

If these equations are rewritten in vector form, the result is

$$\begin{Bmatrix}\alpha_1\\ \alpha_2\\ \alpha_3\\ \alpha_4\\ \alpha_5\\ \alpha_6\\ \alpha_7\\ \alpha_8\\ \alpha_9\end{Bmatrix} = \begin{Bmatrix}0\\ -1\\ 1\\ 0\\ 0\\ 0\\ 0\\ 0\\ 0\end{Bmatrix} c_3 + \begin{Bmatrix}0\\ 0\\ 0\\ -1\\ 1\\ 0\\ 0\\ 0\\ 0\end{Bmatrix} c_5 + \begin{Bmatrix}-1\\ 3\\ 0\\ -2\\ 0\\ 1\\ 0\\ 0\\ 0\end{Bmatrix} c_6 + \begin{Bmatrix}-1\\ 3\\ 0\\ -2\\ 0\\ 0\\ 1\\ 0\\ 0\end{Bmatrix} c_7 + \begin{Bmatrix}-1\\ 3\\ 0\\ -2\\ 0\\ 0\\ 0\\ 1\\ 0\end{Bmatrix} c_8 + \begin{Bmatrix}0\\ -1\\ 0\\ 1\\ 0\\ 0\\ 0\\ 0\\ 1\end{Bmatrix} c_9 \tag{7.11}$$

The columns of Eq. (7.11) represent a separate invariant. The invariant is found by associating each dimensioned quantity with its corresponding α value and raising the dimensioned quantity to the power seen in the column vector. In this case, the invariants are given by

$$\Pi = \left(\frac{d}{l}\right)^{c_3} \left(\frac{u_y}{u_x}\right)^{c_5} \left(\frac{\sigma_{y,c} l^3}{m u_x^2}\right)^{c_6} \left(\frac{E_m l^3}{m u_x^2}\right)^{c_7} \left(\frac{G_o l^3}{m u_x^2}\right)^{c_8} \left(\frac{t u_x}{l}\right)^{c_9} \tag{7.12}$$

The separate invariants are found by setting one $c_i = 1$ for a given i and the others to zero, which gives

$$\begin{aligned}&\pi_1 = \frac{d}{l}, \quad \pi_2 = \frac{u_y}{u_x}, \quad \pi_3 = \frac{\sigma_{y,c} l^3}{m u_x^2}, \\ &\pi_4 = \frac{E_m l^3}{m u_x^2}, \quad \pi_5 = \frac{G_o l^3}{m u_x^2}, \quad \pi_6 = \frac{t u_x}{l}\end{aligned} \tag{7.13}$$

To maintain proper scaling, these six parameters must be matched in between the HHSTT sled and the developed laboratory hypervelocity impact scenario. Term

Table 7.2 HHSTT dimensioned quantities [2, 3]

Dimensioned quantity	Value
Mass m	19.1 kg
Height d	2.54 cm
Length l	20.32 cm
Horizontal velocity u_x	1500 m/s
Vertical velocity u_y	−1 m/s
Compressive yield strength $\sigma_{y,c}$	14.47 GPa
Elastic modulus E_0	180.7 GPa
Shear modulus G_0	70.42 GPa

π_1 of Eq. (7.13) defines the geometry aspect ratio, and π_2 defines the impact angle. Terms π_3 through π_5 relate material properties, length, mass, and horizontal velocity. Parameter π_6 is a timescale that can be used to compare two scenarios.

To scale the HHSTT sled problem, we begin with the known parameters from that gouging scenario. A nominal sled has a mass of 800 kg. The shoes that connect the sled to the rails are generally 20.32 cm long, by 10.8 cm wide, by 2.54 cm high. Taking a unit slice of the geometry for a plane-strain implementation (again, discussed later), the parameters for the HHSTT become those listed in Table 7.2.

Using these values for the HHSTT, the invariant parameters from the Buckingham pi theorem can be computed. Those parameters can then be used in determining the required geometry of a scaled hypervelocity projectile. The π parameters establish ratios between these characteristics. Therefore, to arrive at a design for the laboratory test, we must constrain the solution space with real-world test limitations. Initially, we chose 0.6 cm as the hypervelocity projectile height, which allowed us to generate the remaining values. Unfortunately, this created a scenario that exceeded the capability of the guns available. Consequently, we optimized the available design variables (based on gun limitations) to arrive at the test geometry that best matched the π parameters. The impact angle selected was one that would guarantee gouging because of a limited number of tests available to the authors (discussed in the next section). Table 7.3 summarizes the parameters

Table 7.3 HHSTT dimensioned quantities [2, 3]

Parameter	HHSTT scenario	Theoretical scaled scenario	Actual scaled scenario
π_1	0.125	0.125	0.22
π_2	$-6.67 \cdot 10^{-4}$	$-6.67 \cdot 10^{-4}$	−0.1763
π_3	0.282	0.282	0.9851
π_4	35.619	35.619	125.4
π_5	13.984	13.984	48.88
Horizontal velocity	1500 m/s	4809 m/s	2190 m/s
Vertical velocity	−1 m/s	−3.2 m/s	−386 m/s

that result from the HHSTT geometry and the required scaled parameters for the laboratory tests. Parameter π_6 is not presented, as it is used to compare timescales and not geometry.

If strict adherence to the Buckingham pi approach were possible, the test geometry would have been characterized according to the theoretical column in Table 7.3. However, the gun arrangement available restricted the geometry to projectiles of 4.78 g mass, 5.5 mm diam, 25 mm long cylinders with hemispherical noses. The projectiles were limited because of a requirement to be aerodynamically stable in flight and sized for launch by the specific gun hardware. The maximum velocity achievable from our facility was 2225 m/s. As noted earlier, this limitation prompted an increase in the impact angle in order to ensure sufficient energy was directed into the target rail to generate gouging (discussed in detail in the next section).

Therefore, the theoretical scaled impact test was not possible to conduct given equipment limitations. Another approach needed to be developed to ensure that we could create hypervelocity gouging impacts in the laboratory environment, given a launch velocity limit of 2225 m/s and the projectile limitations just outlined. A one-dimensional penetration theory was adapted to the gouging impact scenario to establish the test parameters. Using this approach, it was possible to create a laboratory gouging test that recreated the major characteristics of hypervelocity gouging. These laboratory gouges could then be used to validate the material flow model and CTH's ability to accurately simulate a hypervelocity gouging impact.

III. One-Dimensional Penetration Model

To better understand the gouging process and to predict when gouging might occur in a scaled laboratory hypervelocity gouging experiment, a one-dimensional penetration theory is refined for use in this particular geometry [13].

A. Theoretical Foundations of the One-Dimensional Penetration Model

Cinnamon and coworkers [14–18] developed a one-dimensional approach to predicting penetration depth and crater diameter based on previous analysis in [19–22]. This theory was based on the penetrator being a rod of known geometry impacting a semi-infinite target material. This theory has application in the HHSTT hypervelocity gouging problem in that the gouging process is considered to begin with rail damage from vertical impact (see Chapter 2). Therefore, we reexamine the already mentioned one-dimensional model and refine its presentation to be utilized as a design tool for a material combination that has no empirical data. That is, this one-dimensional theory can predict penetration depth (rail damage) based on known quantities in the HHSTT gouging problem or define a threshold impact velocity beyond which damage to the rail occurs. The analysis in [18] is reformulated here for clarity.

The general rod penetration process is detailed in Fig. 7.1. An undeformed rod of known geometry (length L and cross-sectional area A_i) impacts a semi-infinite target at a known impact velocity v_0. As the impact event unfolds, the head of the penetrator mushrooms into the target, and material is ejected from the resulting hole; the undeformed section length is ℓ. When the event is complete, a measurable hole remains in the target material, with penetration depth z.

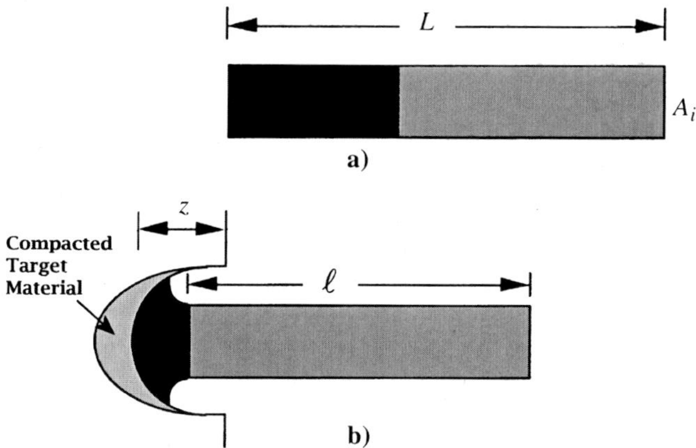

Fig. 7.1 Rod penetration event, a) initial rod geometry, with a shaded region that will be lost to erosion, and b) penetration event.

Jones et al. [19] developed the equation of motion of the undeformed section of the rod (by performing a momentum balance) as

$$\ell\dot{v} + \dot{\ell}(v - u) = \frac{-P}{\rho(1 + e)} \tag{7.14}$$

where the dots refer to differentiation with respect to time, ℓ is the undeformed section length, v is the current undeformed section velocity, u is the penetration velocity, P is the average pressure on the penetrator tip, ρ is the penetrator density, and e is the engineering strain of the penetrator head.

By applying the conservation of mass across the plastic interface between the undeformed section and the mushroom, another relation is determined:

$$e\dot{\ell} = v - u \tag{7.15}$$

Because the engineering strain e in the mushroom is compressive and therefore negative,

$$e = \frac{A_i}{A} - 1 \tag{7.16}$$

where A is the instantaneous penetrator tip cross-sectional area.

This one-dimensional analysis can be improved by adding an initial transient phase, which is dominated by shock effects and complete mushroom growth. This phase precedes a steady-state penetration phase in which further penetrator tip growth is not experienced. This addition was motivated by the observations of Ravid et al. [23].

In applying this transient to the analysis, we assume that the undeformed section length does not experience appreciable deceleration (i.e., $\dot{v} \approx 0$). Therefore, $v = v_0$ during the mushrooming phase of penetration. During this transient, the initial cross-sectional area of the penetrator tip grows from A_0 to A_1. This cross-sectional area A_1 at the end of the transient is maintained for the remainder of the penetration event (again, prompted by the experimental observations in [23]).

Ravid et al. [23] also reported little change in the penetration velocity during this initial transient phase, which motivates the assumption that $u = u_0$, and hence Eq. (7.14) becomes

$$\dot{\ell}(v_0 - u_0) = \frac{-P}{\rho(1+e)} \tag{7.17}$$

and Eq. (7.15) becomes

$$e\dot{\ell} = v_0 - u_0 \tag{7.18}$$

Equations (7.17) and (7.18) describe the mushrooming rod during the initial transient phase. We can arrive at an explicit expression for engineering strain by combining these equations.

$$e = \frac{-(v_0 - u_0)^2}{(v_0 - u_0)^2 + P/\rho} \tag{7.19}$$

This equation describes the strain during the initial transient, until the steady-state penetration phase begins. At that point, the strain is fixed for the remainder of the event.

Once steady-state penetration is reached, the well-known modified Bernoulli equation is applied [19, 22]. This equation relates the pressure on the axis of the penetrator tip p_a, the undeformed section velocity v, the penetration velocity u, and the material properties of the target and penetrator,

$$p_a = \frac{1}{2}\rho_t u^2 + R_t = \frac{1}{2}\rho(v - u)^2 + Y_p \tag{7.20}$$

where ρ_t is the target density, R_t is the target dynamic yield strength, and Y_p is the penetrator dynamic yield strength. We adopt a more explicit formulation of this equation in this work, contrasted to [18], for improved clarity. Making this relationship specific to our transient penetration analysis, Eq. (7.20) becomes

$$p_1 = \frac{1}{2}\rho_t u_0^2 + R_t = \frac{1}{2}\rho(v_0 - u_0)^2 + Y_p \tag{7.21}$$

where p_1 is the axial penetrator tip pressure at the end of the transient. Similarly, Eq. (7.19) becomes

$$e_1 = \frac{-(v_0 - u_0)^2}{(v_0 - u_0)^2 + P_1/\rho} \tag{7.22}$$

where e_1 is the engineering strain and P_1 is the average pressure on the penetrator tip, both quantities taken to be after the transient phase.

Equation (7.21) can be manipulated to solve for u_0 in terms of known quantities in the impact scenario (v_0, ρ, ρ_t, R_t, and Y_p). With a quantity for u_0, Eq. (7.22) can be solved with only the additional quantity P_1. It is the determination of P_1 that is the foundation of this one-dimensional approach.

For the case that the target and penetrator have the same densities ($\rho = \rho_t$) and dynamic yield strengths ($Y_p = R_t$), Eq. (7.21) algebraically reduces to

$$u_0 = \frac{1}{2} v_0 \tag{7.23}$$

This approximation of u_0 also applies for cases in which the impact velocities are relatively small [19].

For unequal dynamic yield strengths ($Y_p \neq R_t$) and equal densities ($\rho = \rho_t$), Eq. (7.21) reduces to

$$u_0 = \frac{\rho v_0^2 + 2(Y_p - R_t)}{2\rho v_0} \tag{7.24}$$

The general penetration case ($Y_p \neq R_t$, $\rho \neq \rho_t$) is given by

$$u_0 = \frac{-\rho v_0}{\rho_t - \rho} + \frac{1}{\rho_t - \rho}\left[\rho^2 v_0^2 - 2(\rho_t - \rho)(R_t - Y_p - \frac{1}{2}\rho v_0^2)\right]^{\frac{1}{2}} \tag{7.25}$$

Therefore, e_1 in Eq. (7.22) is a function of known parameters and the quantity P_1.

At this point, we can also solve for some of the conditions at impact. For instance, the strain at impact e_0 can be computed from Eq. (7.19) if we know the average pressure on the penetrator tip at impact P_0.

$$e_0 = \frac{-(v_0 - u_0)^2}{(v_0 - u_0)^2 + P_0/\rho} \tag{7.26}$$

The impact pressure on the penetrator tip axis p_0 can be estimated from elementary shock physics relationships [24, 25]:

$$p_0 = \rho u_s u_0 \tag{7.27}$$

where u_s in the shock speed in the target. Values for u_s as a function of u_0 can be found in shock Hugoniot tables, for example, [26]. The presence of shock waves in these kinds of high energy impacts has been confirmed by various investigators in the field (for instance, [23]).

Calculation of P_0 from the estimation of p_0 is dependent on the assumed character of the pressure distribution. Previous works have assumed various distributions from constant to highly parabolic (see [17, 18] for a full discussion). In general, P can be computed from

$$P = \frac{1}{A_i}\int_{A_i} p \, \mathrm{d}A_i \tag{7.28}$$

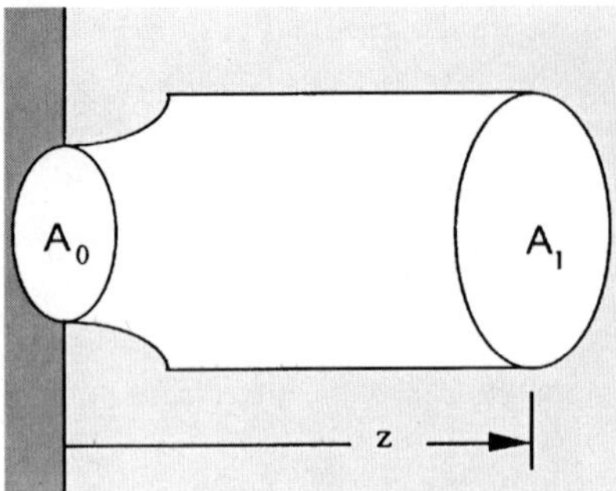

Fig. 7.2 Idealized crater geometry.

A number of pressure profiles were attempted, and it was noted that the more successful modifications to P_1 removed all of the velocity dependence and instead assumed a single constant steady-state average tip pressure for the entire impact velocity range.

The mathematical model for the behavior of the penetrator is a rigid-plastic, instantaneously eroding rod model. As a result, the penetrator enters the target with some impact engineering strain e_0 that expands to e_1 during the transient. The impact pressure p_0 is usually very high relative to the steady state pressure p_1. Although this pressure decreases rapidly during mushroom formation in the transient phase, the values for p_0 can be significant. The mushroom diameter grows from the time of impact through the transient phase and ceases at the beginning of the steady-state portion of the event. The shock/impact stage takes place in a period of a few microseconds [23].

The instantaneous erosion assumption prevents the model from accounting for any additional erosion of the target, which occurs in actual practice. There is typically appreciable change in target geometry caused by penetrator and target material ejection from the crater. As a consequence, the recovered targets will appear to have more cylindrical-type craters than the model would predict. Figure 7.2 illustrates the crater predicted by the mathematical model, and Fig. 7.3 indicates how the actual geometry frequently appears.

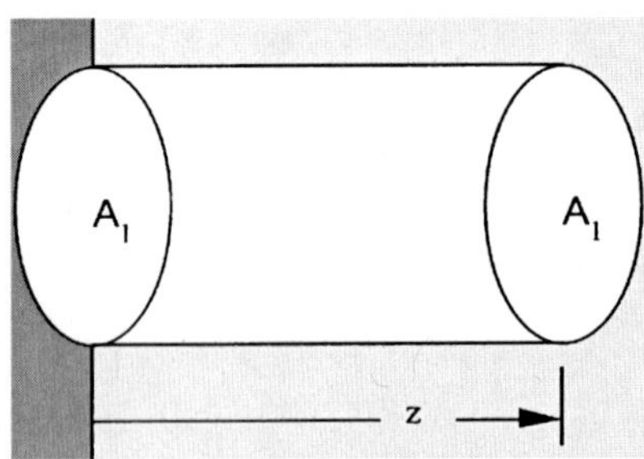

Fig. 7.3 Actual crater geometry.

To this point, the one-dimensional theory can provide an estimate for penetrator strain, given an approximation for P_1. To compute the predicted penetration depth, we need to use the database of empirical test shots. A mathematically sound one-dimensional theory for predicting penetration depth remains elusive.

For a number of years, investigators have observed a very strong correlation between impact kinetic energy and the resulting crater volume. For the velocity range of 1–6 km/s, this relationship is nearly linear [20, 21]. As a result, the crater volume V_c can be simply expressed as

$$V_c = aE_0 + b \tag{7.29}$$

where E_0 is the impact kinetic energy, a is the slope, and b is the intercept of the linear crater volume/kinetic energy relationship. Of course, E_0 can be expressed as

$$E_0 = \frac{1}{2}\rho A_i L v_0^2 \tag{7.30}$$

The linear fit is performed for each shot combination. The results are highly dependent on the quality of the experimental data and sufficient discrete tests.

By utilizing a cylindrical approximation for the crater geometry (as in Fig. 7.3), we can generate an estimate for penetration depths. The cross-sectional area of the target crater will be A_1, which is

$$A_1 = \frac{A_i}{1 + e_1} \tag{7.31}$$

The crater volume then becomes

$$V_c = A_1 z \tag{7.32}$$

We can then predict the penetration depth of an impact scenario by applying the appropriate crater volume/kinetic energy relationship. Combining Eqs. (7.29), (7.31), and (7.32), we get

$$z = \frac{1}{A_i}(1 + e_1)(aE_0 + b) \tag{7.33}$$

Therefore, we have an estimation for the penetration depth based on an empirical crater volume/kinetic energy relationship and the strain at the end of the transient phase, which depends on an approximation of the average pressure on the penetrator tip at steady state P_1.

The determination of P_1 was the focus of previous work. The modified Bernoulli equation tends to overpredict penetration depth significantly as the impact velocities enter the hypervelocity range. A significant effort was made to create pressure profiles that tended to reduce the parabolic nature of the modified Bernoulli relationship [15, 27].

B. Results from the One-Dimensional Penetration Model

Extraordinary results were obtained by disassociating the approximation of P_1 from a particular pressure profile and assuming a constant steady-state penetrator tip pressure for the entire velocity range. That is, the end of the transient and commencement of steady-state penetration occurs at one particular value of average pressure for a specific set of materials. By examining a vast database of existing empirical data [16–18], a strong correlation between the P_1 that best fit the penetration data and the dynamic yield strength of the target was observed. A full discussion of the database used and the limitations of some of the experimental data can be found in [16, 17].

Figures 7.4 and 7.5 summarize the data reduction from [14–18] and report for the first time, in one location, the resulting crater volume/kinetic energy relationship and the estimate for P_1 that resulted in best-fit matches to the empirical penetration depths. That is, estimates for P_1 were chosen for their ability to match experimental data closely. In this process, discovering a correlation between these "best-fit" choices of P_1 and physical parameters in the problem was a priority. It became

Penetrator Material	Y_p (MPa)	ρ (kg/m^3)	L (mm)	L/D	Target Material	R_t (MPa)	ρ_t (kg/m^3)	a (thousands, mm^3/kJ)	b (thousands, mm^3)	P_1 (MPa)
Alum Alloy	200	2700	63.5	10	Lead	200	11200	1 28233	-1 405	128
1100-O Al	250	2720	9.525	3	1100-O Al	250	2720	2.67727	-.227	280
2024-T3 Al	675	2770	9.525	3	1100-O Al	250	2720	2.27238	- 006	280
7075-T6 Al	600	2804	9.525	3	1100-O Al	250	2720	2.69559	- 197	280
C1015 St.	600	7600	9.525	3	1100-O Al	250	2720	5.86367	- 174	280
Soft 4340	1263	7850	31.75	5	2024-T4 Al	400	2770	3 02053	-3.545	600
7075-T6 Al	600	2804	31.75	5	2024-T4 Al	400	2770	97863	- 629	600
Hard 4340	1826	7850	31.75	5	7075-T6 Al	600	2804	1.861	-3.285	1050
Soft 4340	1263	7850	31.75	5	7075-T6 Al	600	2804	1 58951	-2 531	1050
1100-O Al	250	2720	9.525	3	C1015 St.	600	7600	17751	- 046	1050
2024-T3 Al	675	2770	9.525	3	C1015 St.	600	7600	185	- 051	1050
7075-T6 Al	600	2804	9.525	3	C1015 St.	600	7600	18565	- 048	1050
C1015 St.	600	7600	9.525	3	C1015 St.	600	7600	.486	- 048	1050
4340 Steel	1600	7810	68.58	18	6061-T651 Al	600	2710	1.8191	-3 801	1050
1100-O Al	250	2720	9.525	3	2024-T3 Al	675	2770	.79216	- 069	1300
1100-O Al	250	2720	9.525	3	304 St.St.	675	7900	14375	- 028	1300
2024-T3 Al	675	2770	9.525	3	2024-T3 Al	675	2770	.77581	-.051	1300
2024-T3 Al	675	2770	9.525	3	304 St.St.	675	7900	15862	- 052	1300
C1015 St.	600	7600	9.525	3	2024-T3 Al	675	2770	1 37937	- 043	1300
C1015 St.	600	7600	9.525	3	304 St.St.	675	7900	34775	- 01	1300
304 St.St.	675	7900	9.525	3	2024-T3 Al	675	2770	1.91916	-.24	1300
304 St.St.	675	7900	9.525	3	304 St.St.	675	7900	.43117	- 046	1300
C110W1	1200	7850	25	10	St37	750	7850	.4056	- 151	1400
C110W1	1200	7850	43	10	St37	750	7850	.39881	-.85	1400
C110W1	1200	7850	54	10	St37	750	7850	.4573	-1.373	1400
C110W1	1200	7850	25	10	St52	850	7850	.32086	-.088	1700
C110W1	1200	7850	43	10	St52	850	7850	.33567	- 448	1700
C110W1	1200	7850	54	10	St52	850	7850	.30658	-.674	1700
D17	2500	17000	28	10	St52	850	7850	.73457	-.291	1700
D17	2500	17000	60	10	St52	850	7850	72127	-1.812	1700
C110W2	1100	7850	58	10	HzB,A	850	7850	.30214	-1.427	1700
Marag St	1000	7850	58	10	HzB,A	850	7850	.55887	-3 28	1700
Marag St	1000	7850	116	20	HzB,A	850	7850	.93885	-18.371	1700
35CrNiMo	2200	7850	54	10	HzB,A	850	7850	.33035	-2.768	1700
Elmet	2000	15500	58	10	HzB,A	850	7850	.51019	-1 189	1700
D17K	2500	17300	58	10	HzB,A	850	7850	.54269	-1.58	1700
D17	2500	17000	58	10	HzB,A	850	7850	.57631	-1 794	1700
W	2500	19300	60	10	HzB,A	850	7850	.67001	-2 742	1700
W75	2500	15500	58	10	HzB,A	850	7850	49547	-1.264	1700
W90	2500	17000	58	10	HzB,A	850	7850	.55021	1.470	1700
D18	2500	18000	58	10	HzB,A	850	7850	.55090	-1.403	1700
H01T	2000	14500	58	10	HzB,A	850	7850	31125	-.048	1700
D17.6	2000	17600	41.7	10	HzB,A	850	7850	.59808	-1.722	1700
D17.6	2000	17600	58	10	HzB,A	850	7850	.64719	-2.775	1700
Steel	1600	7850	58	10	HzB,A	850	7850	.1301	-.009	1700
D17	2500	17000	116	20	HzB,A	850	7850	47207	-3.142	1700

Fig. 7.4 Empirical data summary, part 1.

Penetrator Material	Y_p (MPa)	ρ (kg/m^3)	L (mm)	L/D	Target Material	R_t (MPa)	ρ_t (kg/m^3)	a (thousands, mm^3/kJ)	b (thousands, mm^3)	P_1 (MPa)
D18.5	2500	18500	58	10	HzB,A	850	7850	.50749	797	1700
H01T	2000	14500	58	10	HzB,A	850	7850	.5255	-2 544	1700
H60T	1600	13500	60	10	HzB,A	850	7850	.45625	-1.153	1700
H60T	1600	13500	58	10	HzB,A	850	7850	.53628	-2.431	1700
H70T	1400	13500	60	10	HzB,A	850	7850	.50427	-1.696	1700
H70T	1400	13500	58	10	HzB,A	850	7850	.51743	-1 364	1700
D17.6	2000	17600	101.5	17.5	HzB,A	850	7850	.743	-9.305	1700
D17.6	2000	17600	107.8	22	HzB,A	850	7850	.887	-16.427	1700
D17.6	2000	17600	110.25	22.5	HzB,A	850	7850	.35615	-1.552	1700
D17.6	2000	17600	156.8	32	HzB,A	850	7850	.36932	-2.853	1700
D17.6	2000	17600	163.2	32	HzB,A	850	7850	.90932	-19.938	1700
Hard 4340	1826	7850	60	6	RHA	1000	7850	.33074	-12.832	2200
OFHC Cu	300	8900	60	6	RHA	1000	7850	.44	-12.862	2200
Tantalum	500	16600	60	6	RHA	1000	7850	.59013	-6.425	2200
Hard 4340	1826	7850	95.25	7.5	RHA	1000	7850	.21294	-8.144	2200
Kenn W10	2500	17300	155.8	23	RHA	1000	7850	.60321	-23 912	2200
Kenn W10	2500	17300	121.75	23	RHA	1000	7850	57257	-13.244	2200
D17	2500	17000	28	10	W8	1000	7850	.49159	-.288	2200
U-3/4Ti	7000	18600	266.7	20	RHA	1000	7850	.54275	-165.234	2200
Kenn W10	2500	17200	50	10	Ger RHA	1000	7850	.47692	-1.299	2200
Teledy X27	2500	17330	78.74	10	RHA	1000	7850	50125	-5 403	2200
C110W1	1200	7850	25	10	Ger Arm St	1100	7850	.22431	-.094	2275
C110W1	1200	7850	43	10	Ger Arm St	1100	7850	.22444	- 359	2275
C110W1	1200	7850	54	10	Ger Arm St	1100	7850	.24157	-.97	2275
D17	2500	17000	28	10	Ger Arm St	1100	7850	.52198	-.335	2275
D17	2500	17000	60	10	Ger Arm St	1100	7850	.39291	- 897	2275
D17.6	2000	17600	17.4	3	Ger St	1200	7850	.57051	-1.474	2350
D17.6	2000	17600	29	5	Ger St	1200	7850	.58488	-1.431	2350
D17.6	2000	17600	42	10	Ger St	1200	7850	.86399	-7.527	2350
Soft 4340	1263	7850	31.75	5	Soft 4340	1263	7850	.2248	- 004	2400
7075-T6 Al	600	2804	31.75	5	Soft 4340	1263	7850	.12627	- 764	2400
Hard 4340	1826	7850	31.75	5	Soft 4340	1263	7850	.30299	- 884	2400
X21C	3500	17650	45.7	10	4340 Steel	1600	7810	.41786	-1.08	2500
Teledy X27C	2500	17400	81.6	10	4340 Steel	1600	7810	.35178	-1.915	2500
Teledy X27C	2500	17400	81.6	15	4340 Steel	1600	7810	.31052	-1.249	2500
Hard 4340	1826	7850	63.5	10	Hard 4340	1826	7850	.17261	-.562	2600
Hard 4340	1826	7850	47.63	7.5	Hard 4340	1826	7850	.16663	- 379	2600
Hard 4340	1826	7850	47.63	5	Hard 4340	1826	7850	.26591	-2.524	2600
Hard 4340	1826	7850	31.75	5	Hard 4340	1826	7850	.22845	- 73	2600
Hard 4340	1826	7850	63.5	5	Hard 4340	1826	7850	.24281	-3.068	2600
Hard 4340	1826	7850	31.75	3.33	Hard 4340	1826	7850	.27563	-1.211	2600
Hard 4340	1826	7850	31.75	2.5	Hard 4340	1826	7850	.25713	-1.228	2600
Soft 4340	1263	7850	31.75	5	Hard 4340	1826	7850	.26557	-.841	2600
OFHC Cu	300	8900	60	6	Hard 4340	1826	7850	.26425	-3.517	2600
Tantalum	500	16600	60	6	Hard 4340	1826	7850	.39113	2.931	2600
D17	2500	17000	28	10	D17	2500	17000	.22515	-.224	2900

Fig. 7.5 Empirical data summary, part 2.

clear that a strong correlation exists between the approximation of P_1 and the dynamic yield strength of the target. Figure 7.6 illustrates this correlation. The cases in which figures appear in this text are annotated by color highlighting around the entries.

The fit in Fig. 7.6 can be expressed as

$$P_1 = 3.8\left[1 - e^{(-0.00135R_t)}\right] - 0.8 \tag{7.34}$$

with P_1 expressed in GPa and R_t expressed in units of MPa in this instance.

Figures 7.7 and 7.8 illustrate the improvements this revised average pressure estimate makes in both the estimate of strain and penetration depths for a typical material combination, Kennenmetal W10 on rolled homogenous armor (RHA). This combination is highlighted in gray in Fig. 7.5. Figures 7.9–7.13 provide two additional examples to those presented in [18] to illustrate the capability of this theory to predict target damage. These cases are highlighted in red and blue in Fig. 7.4. Figures 7.11–7.13 are chosen specifically to apply to the hypervelocity gouging scenario, discussed later.

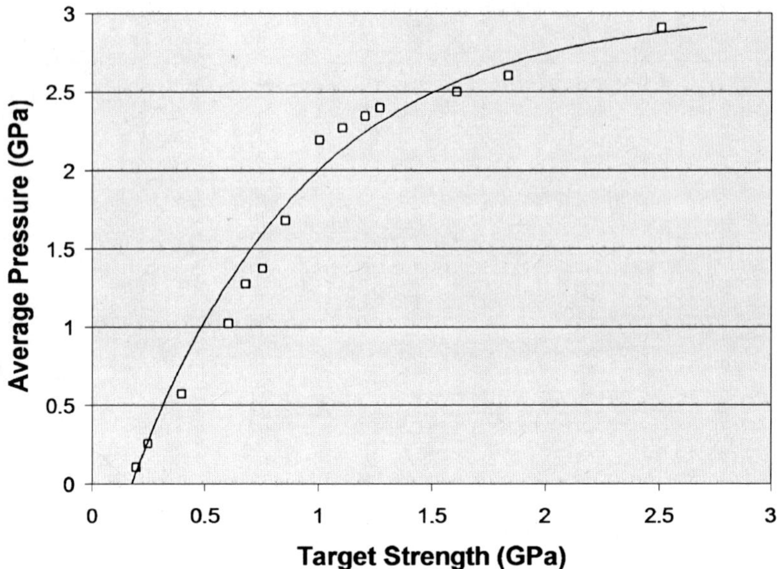

Fig. 7.6 P_1 vs target dynamic yield strength.

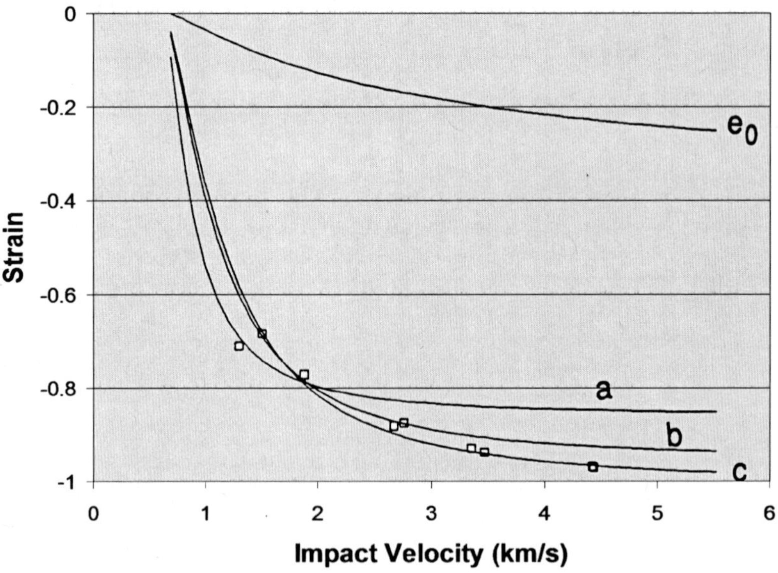

Fig. 7.7 Strain comparison of average pressure estimates, Kennenmetal W10 on RHA: a) from approach in [14, 15], b) from approach in [27], and c) from approach in [13, 15, 16, 18].

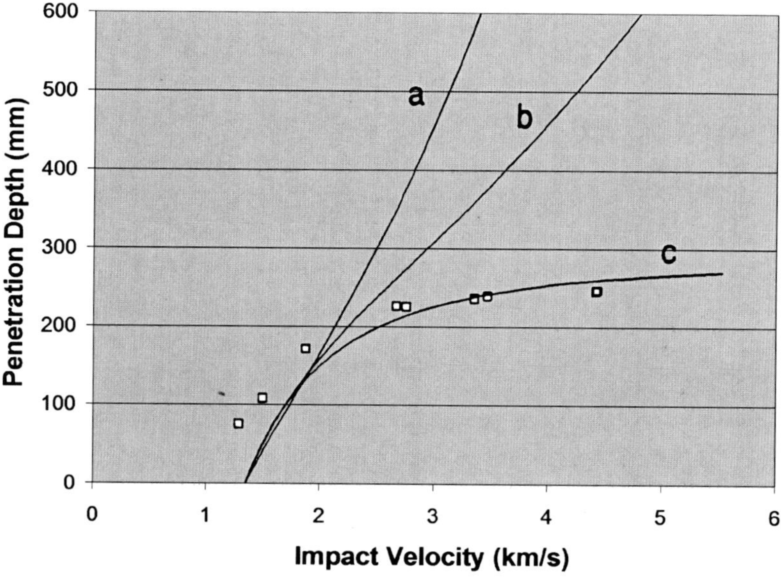

Fig. 7.8 Penetration comparison of average pressure estimates, Kennenmetal W10 on RHA: a) from approach in [14, 15], b) from approach in [27], and c) from approach in [13, 15, 16, 18].

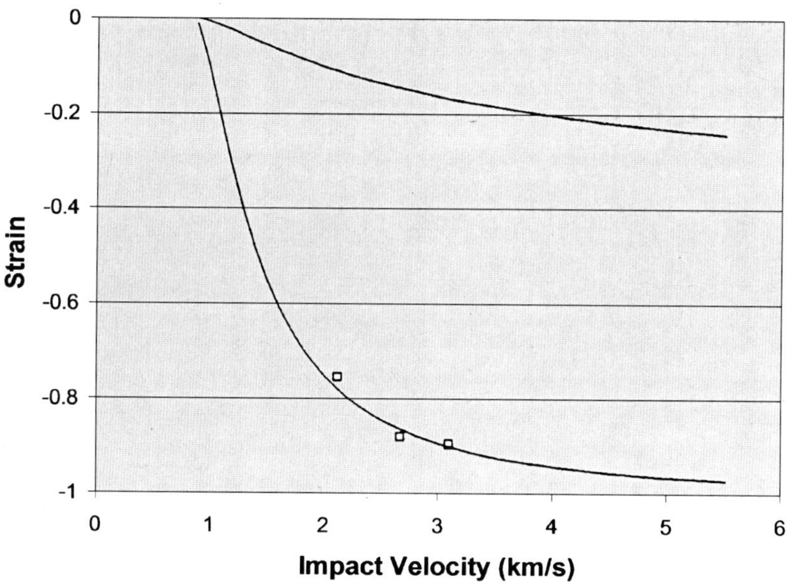

Fig. 7.9 Strain, 4340 steel on 6061-T651 aluminum.

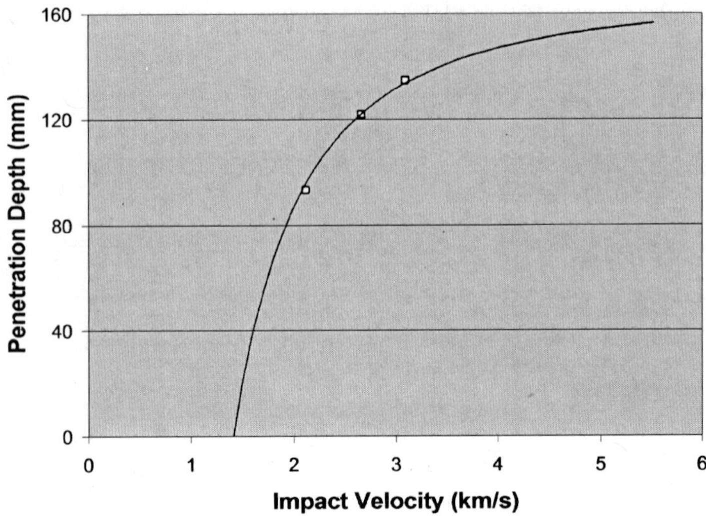

Fig. 7.10 Penetration depth, 4340 steel on 6061-T651 aluminum.

C. Engineering Design Approach for Using One-Dimensional Penetration Model

One of the driving motivations for the development of this one-dimensional approximation is to develop an algorithm for engineering design efforts. That is, we wish to establish a database and a procedure for an estimation for target damage without the need to resort to expensive experimentation or time-intensive

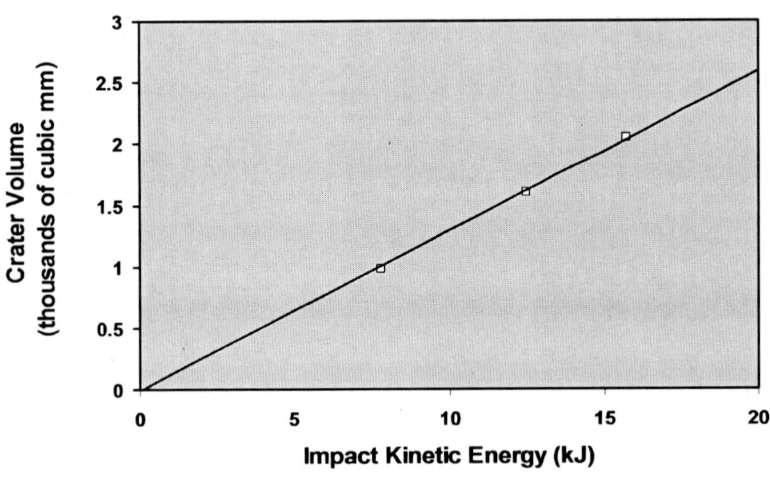

Fig. 7.11 Crater volume/kinetic energy, steel on HzB, A.

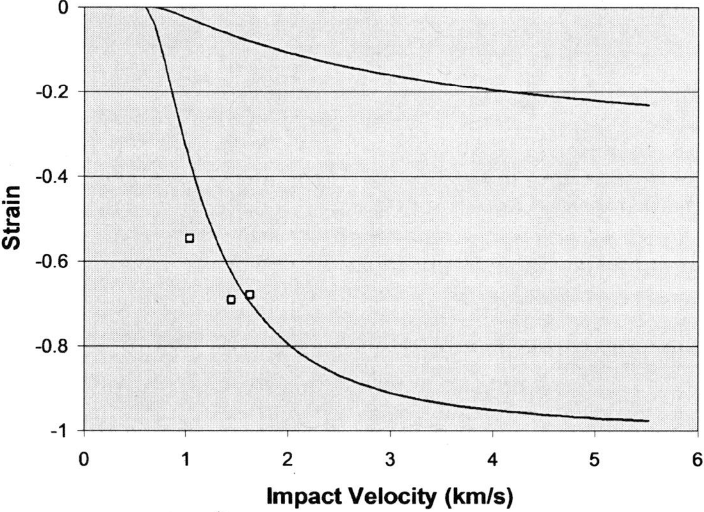

Fig. 7.12 Strain, steel on HzB, A.

computational simulations. Of course, this would only serve as a first-order approximation. However, it promises to provide a good approximation prior to code simulations or gun range work.

The one-dimensional approach for predicting penetration depth can be used to make a first-order approximation in impact scenarios that have not been experimentally performed. To establish this capability, we note that the modified Bernoulli equation and the theory developed rely on a few specific material parameters

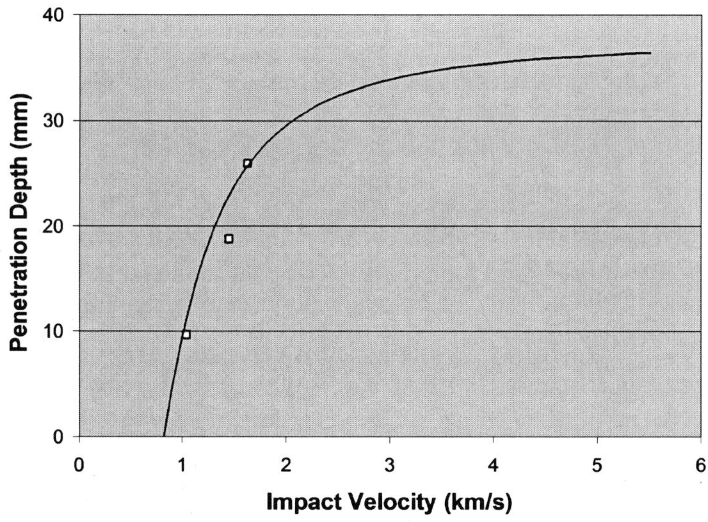

Fig. 7.13 Penetration depth, steel on HzB, A.

(i.e., dynamic yield strength and density). If the materials to be evaluated can be found to closely match those presented in Figs. 7.4 and 7.5, a simple algorithm to estimate the resulting damage to a target given a set of initial impact conditions can be followed.

1) Select a material combination in Figs. 7.4 and 7.5 that matches fairly closely the density and dynamic yield strengths of a scenario to be evaluated. The appropriate crater volume/kinetic energy relationship is then selected. Note that a detailed discussion on estimating the dynamic yield strength is available elsewhere. Estimates for this flow stress are typically gained from other high strain-rate experimentation (e.g., split Hopkinson bar, flyer plate, Taylor impact tests, etc.).

2) Select the appropriate expression for u_0 from Eqs. (7.23–7.25) and compute the value. This value can then be inserted into Eq. (7.22) with the appropriate value of P_1 from Figs. 7.4 and 7.5 to arrive at e_1. In cases in which the estimated dynamic yield strength of the target material is not sufficiently close to the value reflected in Figs. 7.4 and 7.5, P_1 can be computed from the relationship in Eq. (7.34).

3) The crater cross-sectional area and the penetration depth can then be computed from Eqs. (7.31) and (7.33).

In this manner, a simple one-dimensional approximation for the penetration depth and crater diameter can be made using known quantities of a hypothetical scenario and the empirical relationships outlined in Figs. 7.4 and 7.5.

For our specific application, we can use this simple approximation to estimate the required vertical velocity to damage the HHSTT rail and initiate gouging or work from known rail damage to compute the velocities required for that damage to occur. Therefore, this penetration theory can be used to determine required laboratory configuration to ensure gouging, given the equipment capabilities. Additionally, this theory can be applied to the full HHSTT problem to serve as an analytical modeling tool that does not require significant computational resources to utilize. This application is presented in the next section.

To verify that this one-dimensional theory is valid in our laboratory gouging environment, we can compare the theory results to known gouging parameters from the HHSTT problem. Successful estimation of those conditions will increase our confidence in this analytical approach.

D. Application of the One-Dimensional Theory to the HHSTT Gouging Problem

The one-dimensional approach for predicting penetration depth can be used to predict the occurrence of gouging in the hypervelocity gouging impact found at the HHSTT. A great deal of work has been focused on determining a threshold velocity for gouging to occur. Many investigators have concluded that a vertical impact is necessary to initiate the gouging process [28–34]. They argue that an initial vertical deformation of the rail causes an asperity on the rail surface, which initiates the high pressure core and material mixing that characterizes gouging.

The one-dimensional approach just described can be utilized to compute the threshold vertical velocity component that creates a rail penetration and to compare that against known parameters that lead to gouging at the HHSTT. To apply this approach to our hypervelocity gouging problem, we follow the computation steps presented in the preceding section.

The material combination chosen in Figs. 7.11–7.13 match closely with the values for density and dynamic yield strength of 1080 steel and VascoMax 300 found using the split Hopkinson bar test (see Chapter 5). This relationship is highlighted in blue in Fig. 7.4. We then take the threshold penetration kinetic energy from Fig. 7.11 as 69.5 J. Assuming that the nominal sled shoe assumed $\frac{1}{4}$ of the sled mass, a 200-kg mass would need to impact the rail at approximately 0.84 m/s. If we were to assume that some pitch or yaw in the sled had caused a single shoe to assume $\frac{1}{2}$ the sled mass, then a 400-kg mass would need to impact the rail at about 0.59 m/s to begin penetrating the rail. This penetration would be the precursor, but not the guarantee, of gouging.

The values arrived at using this simple one-dimensional analysis are precisely within the range predicted by aerodynamic study and reported by the HHSTT as probable conditions during a hypervelocity test run [35]. Additionally, this magnitude (1–2 m/s) was used in the hydrocode CTH to initiate gouging by Szmerekovsky et al. [9].

Taking this analysis one step further, we can assume that same 400 kg came down at 5 m/s and that the leading 10% of the shoe surface area impacted the rail (resulting from pitch angle, for instance). With nominal shoe dimensions of 20.32×10.8 cm, the resulting penetration depth would be 0.3 mm, which would be sufficient to cause gouging to initiate [9]. In fact, 0.1 mm was shown to be sufficient to initiate the typical gouge event within code simulations reported in [9]. If we further consider the penetrator surface to be 1% of the nominal shoe surface area, the resulting penetration depth would be 3.0 mm. This depth prediction approaches the magnitude of material removal (~5 mm) noted in Chapter 4 from a gouged rail from the HHSTT.

Therefore, this one-dimensional theory does match previous CTH code computations and actual gouges in the field remarkably well and can provide insight into the parameters (such as vertical impact velocity and material dynamic yield strength) that govern hypervelocity gouging.

With this level of validation established for the one-dimensional penetration theory approach, the theory was used to establish the test parameters for the scaled laboratory hypervelocity gouging tests.

E. Application of the One-Dimensional Theory to the Laboratory Hypervelocity Gouging Tests

Using the same approach as described in the preceding section, the one-dimensional theory was applied to the estimation of the experimental configuration for the laboratory hypervelocity gouging tests. Recall that the test limitations were a projectile of approximately 5 g in mass, 5.5 mm in diameter, and 25 mm in length. Applying this dimensionality to the equations outlined in the preceding section, the threshold penetration velocity yielded a striking angle of about 4.4 deg. If we use the previously reported value for minimum gouge depth in CTH models to create gouging of 0.1 mm, the angle would need to be approximately 8 deg.

Therefore, striking angles of 10 and 15 deg were chosen for the experimental tests. The critical nature of having these estimates prior to testing was that a small number of tests (four) was available. The tests were then conducted in order to generate hypervelocity gouging in the laboratory sense.

IV. Laboratory Hypervelocity Gouging Experiments and Validation of the One-Dimensional Penetration Model

Based on the Buckingham pi analysis presented earlier in this chapter, and constrained by the capability of our laboratory gun facility, the test parameters for a scaled hypervelocity gouging test were chosen. Although matching the parameters exactly was not possible, an attempt was made to establish a test geometry that reduced the difference between the theoretical parameters and the experimental parameters. The goal, of course, was to generate hypervelocity gouging within a laboratory scenario.

The experimental apparatus was available at the 46th Test Wing, Wright-Patterson Air Force Base, Ohio. The projectile was fired by a 30-mm powder gun as pictured in Fig. 7.14. The projectile was held in the barrel by a sectioned sabot that split apart during free flight, as pictured in Fig. 7.15. The sabot sections hit the "sabot stripper" plate, and the projectile traveled through the hole to hit the target rail as shown in Fig. 7.16.

The impact was digitally captured using two high-speed cameras at a frame rate of 47,000 frames per second. An attempt was made to "soft-catch" the projectiles, which proved ultimately to be unsuccessful. The impact velocity was measured using "break-sheets" immediately prior to the target. The electrical current breaks were recorded, and the impact velocity was computed for each shot.

Fig. 7.14 30-mm powder gun.

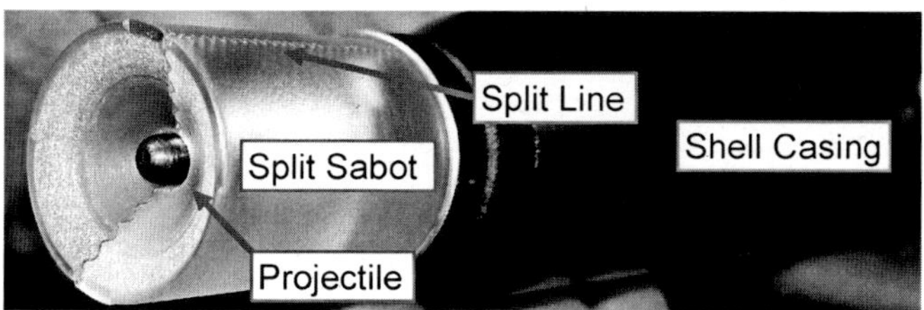

Fig. 7.15 Projectile in sabot.

A series of equipment "check-out" shots were conducted to ensure the functionality of the configuration. For these initial qualification tests, stock steel was chosen for the target and the penetrator (304 stainless steel). Three tests shots were accomplished. The target plate was set at an oblique angle of 10 deg to the penetrator path to create a gouging type impact. Figure 7.17 illustrates a typical gouge generated in the 304 stainless steel (projectile motion was left to right). Table 7.4 summarizes the test results for this qualification test series.

To further validate our one-dimensional model, we chose the 304 stainless steel on 304 stainless steel from Fig. 7.4 (outlined in green). This gives us a crater volume/kinetic energy relationship. We can then compute u_0 and a penetration depth z. This computation is made slightly more difficult because the projectile impacts along its long axis, and therefore the cross-sectional area (in a one-dimensional sense) in contact with the target changes as a function of penetration depth. That is, as penetration depth is computed, more of the cylinder's cross-sectional area is in contact with the target, changing A_i. Therefore, Eq. (7.33) is

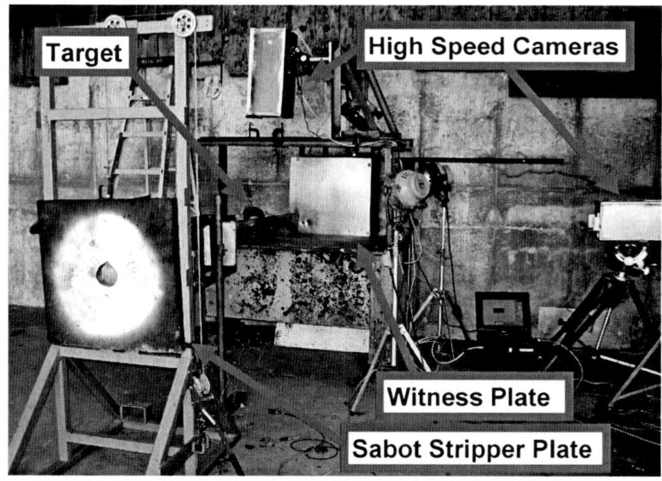

Fig. 7.16 Target area.

Fig. 7.17 Gouged 304 stainless-steel target (test ss-1).

modified by a factor of $4/3\pi$ (to account for the hemispherical shape of the penetrating edge of the projectile), and a simple iteration scheme based on a cross-section calculation and predicted penetration depth converges to the values in Table 7.4. Of course, only the component of the velocity vector oriented into the target (i.e., normal to the target surface) is used to compute the penetration kinetic energy.

The measured crater depth for each shot also appears in Table 7.4. Good agreement between experimentation and the one-dimensional model was observed. Therefore, the general approximation regarding penetration depth was validated in another application.

After the laboratory configuration qualification test shots, the HHSTT materials were shot. During the initial tests of VascoMax 300 projectiles shot against the 1080 steel railroad rails used at the HHSTT, a 7.3-g projectile (5.5 mm diam, 37.5 mm length) was used. These projectiles tended to tumble too much and would not successfully go through the stripper plate hole. However, one of these tests provided an opportunity to validate our one-dimensional model in a more traditional normal impact scenario. One of these larger projectiles impacted the stripper plate with the long axis parallel to the plate surface at 2226 m/s with a 0 deg of incidence. Figure 7.18 is the resulting crater. Table 7.5 summarizes the measured and predicted quantities of this impact event. The crater volume/kinetic energy relationship used is the same for VascoMax 300 on 1080 steel because the stripper plate steel is a close match to 1080 steel.

The predicted values match the measured quantities closely. In fact, taking the computed impact kinetic energy of 18.16 kJ and the measured crater volume, we can add an additional point on the crater volume/kinetic energy relationship presented in Fig. 7.11. The additional experimental point falls on the linear fit we have been using. Figure 7.19 shows this new relationship.

The test series of VascoMax 300 projectiles (4.78 g, 5.5 mm diam, 25 mm length) was then conducted. Based on the predictions of penetration depth available using the theory, angles of 10 and 15 deg were selected to create gouging. The gouging event was recorded via high-speed digital photography. The high-energy/high thermal character of the event was immediately evident [36].

Table 7.4 304 stainless on 304 stainless-steel test series

Test	Angle, deg	Impact velocity, m/s	z (predicted), mm	z (measured), mm
ss-1	10	2584	0.99	0.9 ± 0.1
ss-2	10	2147	0.73	0.7 ± 0.1
ss-1	10	2157	0.73	0.7 ± 0.1

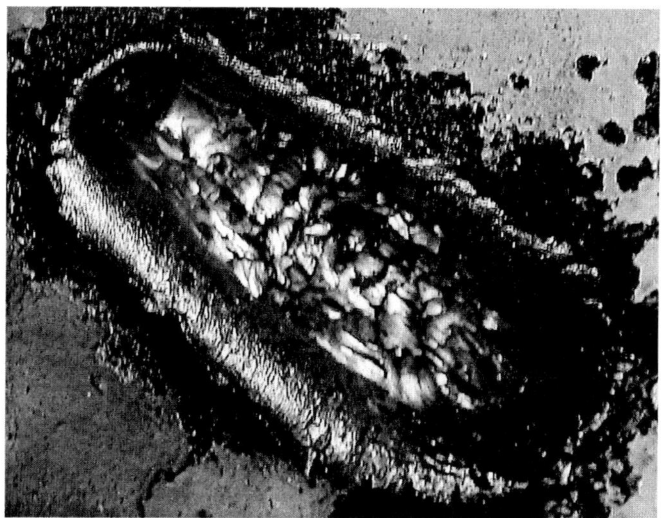

Fig. 7.18 Crater in sabot stripper plate (test sp-1).

Table 7.5 VascoMax 300 impact on steel stripper plate

Parameter (test sp-1)	Predicted	Measured
Penetration depth	2.41 mm	2.5 ± 0.1 mm
Crater volume	2351.8 mm^3	2400 ± 300 mm^3
Crater area	1261.5 mm^2	1200 ± 80 mm^2

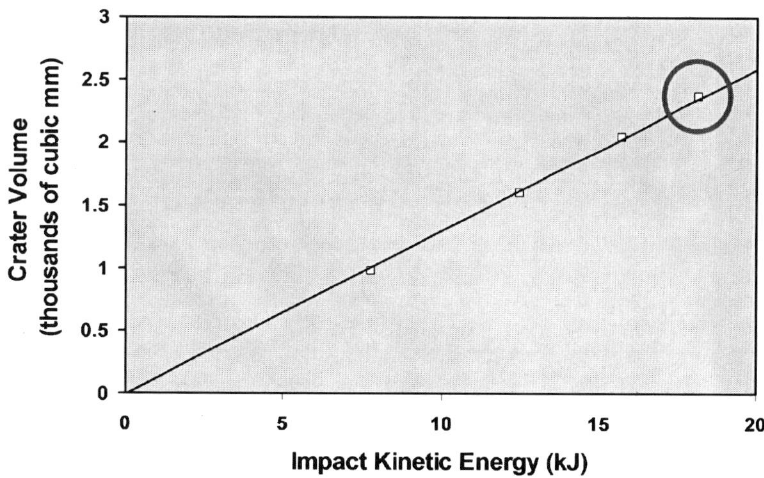

Fig. 7.19 Updated crater volume/kinetic energy, steel on HzB, A.

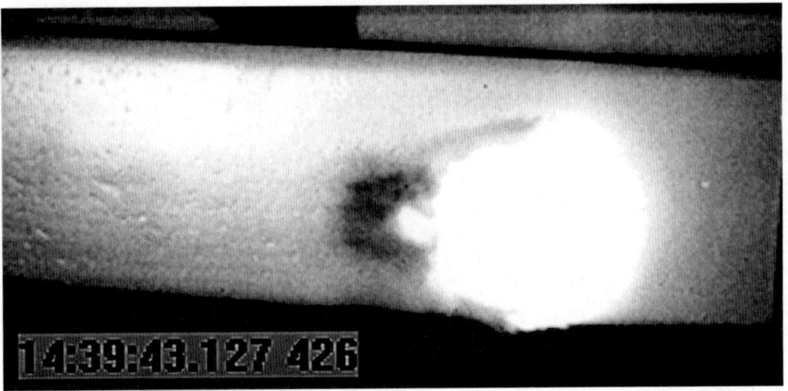

Fig. 7.20 Typical high-speed photo of gouging event.

Figure 7.20 is a typical impact of the projectile against the rail (in this case coated with epoxy), with the impact event itself obscured by a bright fireball. The projectile enters the frame from the left and impacts the rail face proceeding to the right.

Two rail targets were used. One was coated with iron oxide and the other coated with epoxy. These coatings represent the two used at the HHSTT on the rail to mitigate gouging. Figures 7.21 and 7.22 depict typical gouges on these targets.

Table 7.6 summarizes the experimental results and the prediction generated by the one-dimensional theory. Again, close agreement is observed between theory and experiment. There was no measurable difference in the gouges created in the rails coated with different materials.

These results further illustrate the ability of the one-dimensional model to predict penetration depth. Of course, because of the horizontal velocity and the nature of a gouging impact, the crater volume cannot be used to compare against the theory. That is, the gouging creates a much larger crater because of the horizontal velocity of the projectile.

Fig. 7.21 Gouge in iron-oxide-coated rail (test hi-2).

Fig. 7.22 Gouge in epoxy-coated rail (test he-1).

Table 7.6 VascoMax 300 on 1080 steel test series

Test	Coating	Angle, deg	Impact velocity, m/s	z (predicted), mm	z (measured), mm
hi-1	Iron oxide	10	2225	0.51	0.5 ± 0.1
hi-2	Iron oxide	15	2150	1.03	1.0 ± 0.1
he-1	Epoxy	15	2147	1.03	1.0 ± 0.1
he-2	Epoxy	10	2163	0.48	0.5 ± 0.1

In addition to these impacts, a couple of unintended impacts also allowed us to further validate this one-dimensional prediction. To prevent the projectiles from lodging into the sabot material during the explosive launch, a steel "pusher plate" ($10 \times 10 \times 5$-mm, 3.925 g) was constructed and placed at the rear of the projectile. In two of the tests (hi-2, he-1), the pusher plate also impacted the rail (in different locations from the primary gouge). The secondary impacts also gouged the rail and provided us with another opportunity to apply our theory. Figure 7.23 shows one of these pusher plate impacts.

In applying our approach to this impact combination, a material combination that most closely matches the dynamic yield strength and densities of the pusher plate steel and 1080 steel were selected from Fig. 7.4 and outlined in orange. Table 7.7 summarizes the measured and predicted penetration depths from this material combination.

Despite being unintended impacts, these pusher plate gouges also demonstrate agreement between the one-dimensional theory and experimental results. Figure 7.24 summarizes the match between all of the experimental results and the predictions. The uncertainty in test measurement is depicted with the black error bars.

The one-dimensional theory matches experimental results extremely well. The ability of it to match the experimental results, across the various material combinations, illustrates its flexibility. Specifically, the theory has been validated against

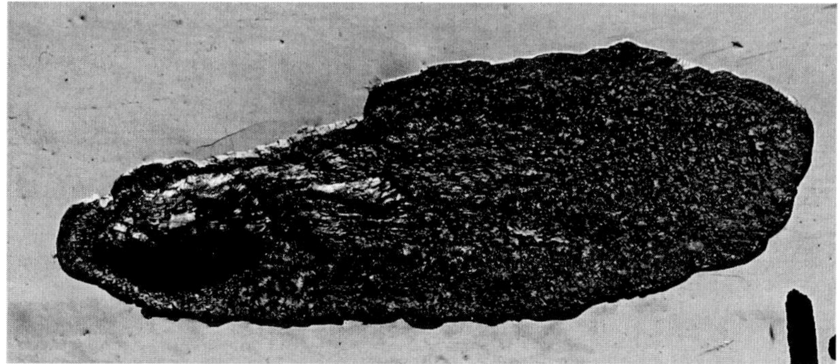

Fig. 7.23 Pusher plate impact on rail (test he-1).

Table 7.7 Pusher plate steel on 1080 steel impacts

Test	Coating	Angle, deg	Impact velocity, m/s	z (predicted), mm	z (measured), mm
pp-1 (from hi-2)	Iron oxide	15	2150	0.91	1.0 ± 0.1
pp-2 (from he-1)	Epoxy	15	2147	0.90	1.0 ± 0.1

experimental shots involving the materials in the HHSTT scenario. This, coupled with the preceding analysis indicating the theory matches previous modeling of gouging and results from the field, confirms that this one-dimensional theory can be a powerful tool in predicting gouging.

As a final experimental examination of the scaled laboratory hypervelocity gouging tests, a metallurgical examination of the gouges was performed.

V. Metallurgical Examination of Hypervelocity Gouging Test Results

The in-depth characterization of gouging presented in Chapter 4 indicated that one of the primary characteristics of the plastic deformation is a discernible heat zone in the posttest specimens. In an effort to verify that the gouging impacts generated with the scaled laboratory hypervelocity gouging tests presented in this chapter generated the same phenomenon, a metallurgical study was performed on the gouged rails.

Utilizing the exact techniques outlined in Chapter 4, three of the four gouges were prepared for microscopy using the as-polished techniques. The samples were cut along the direction of projectile motion, down the middle of the gouge. The rail

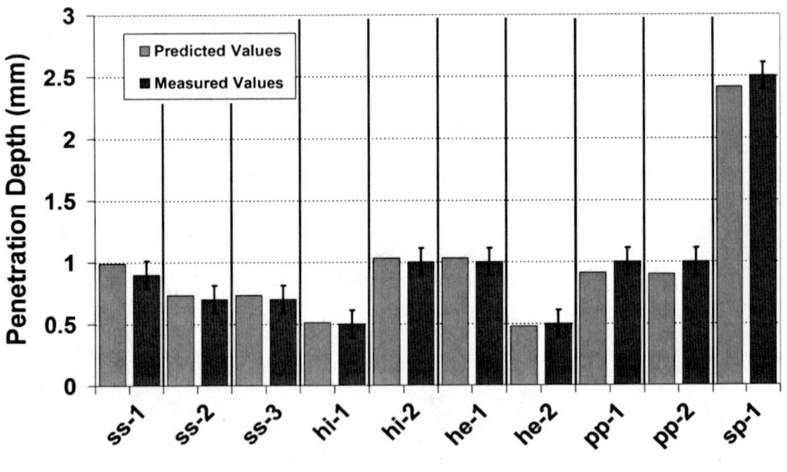

Fig. 7.24 Penetration depth summary.

was then examined through the depth, in the plane parallel to projectile motion (exactly as in Chapter 4).

All of the gouged specimens exhibited the same evidence of a thermal pulse of sufficient magnitude as to austenize the 1080 steel (above 725°C). Figure 7.25 is characteristic of what was discovered.

Fig. 7.25 Metallurgical examination of gouge test he-1.

In this figure, the microstructure of the steel near the gouge surface is modified from the unaffected microstructure shown on the same figure (which is taken at the rail surface in an undamaged section that still has its epoxy coating on it). The microstructure of the gouged section changes back to an unaffected state at about 1.0 mm into the rail depth. The horizontal black line in the gouged section's micrograph indicates a jump to a location far away from the gouge surface (about 2 cm).

It is clear from this examination that the hypervelocity gouge tests conducted in the laboratory in a scaled form produced the same phenomenology as the full-scale gouge from the HHSTT. More detail is provided on this metallurgical analysis in the next chapter.

VI. Summary

A scaled hypervelocity gouging test was developed for a laboratory application. The goal was to create hypervelocity gouging experimentally for the purpose of validating our material constitutive models developed earlier in this work. A mathematical scaling approach was used to determine the experiment's parameters, which exceeded the available gun range capability.

A one-dimensional penetration theory was developed to aid in the determination of experimental parameters for a hypervelocity impact test that was feasible. The one-dimensional approach matched previous CTH models of the HHSTT problem, and data from the field, very well. The one-dimensional theory was then applied to the scaled hypervelocity experiment to set test parameters.

A series of hypervelocity impact tests was conducted. The one-dimensional theory was validated against several different material combinations and predicted the resulting penetration depths extremely well. Hypervelocity gouging impacts were created in a scaled laboratory experiment. These gouges were shown to have the same microstructural characteristics as full-scale gouges from the HHSTT.

Therefore, this approach has yielded hypervelocity gouging impacts that can be modeled by CTH, with the newly developed full-range constitutive models, to validate the codes predictions of these gouging events.

References

[1] Cinnamon, J. D., Palazotto, A. N., Szmerekovsky, A. G., and Pendleton, R. J., "Further Investigation of a Scaled Hypervelocity Gouging Model and Validation of Material Constitutive Models," AIAA Paper 2006-2087, May 2006.

[2] Rickerd, G. S., "An Investigation of a Simplied Gouging Model AFIT/GAE/ENY 05-M19," Master's Thesis, Department of Aeronautics and Astronautics, Air Force Inst. of Technology, Wright-Patterson AFB, Dayton, OH, March 2005.

[3] Rickerd, G. S., Palazotto, A. N., and Cinnamon, J. D., "Investigation of a Simplified Hypervelocity Gouging Model," AIAA Paper 2005-2355, April 2005.

[4] Szmerekovsky, A. G., "The Physical Understanding of the Use of Coatings to Mitigate Hypervelocity Gouging Considering Real Test Sled Dimensions AFIT/DS/ENY 04-06," Ph.D. Dissertation, Department of Aeronautics and Astronautics, Air Force Inst. of Technology, Wright-Patterson AFB, Dayton, OH, Sept. 2004.

[5] Pendleton, R. J., "Validation of a Scaled Plane Strain Hypervelocity Gouging Model AFIT/GAE/ENY/06-M26," Master's Thesis, Department of Aeronautics and Astronautics, Air Force Inst. of Technology, Wright-Patterson AFB, Dayton, OH, March 2006.

[6] Barenblatt, G. I., *Dimensional Analysis*, Gordon and Breach, New York, 1987.

[7] Cengel, Y. A., and Cimbala, J. M., *Fluid Mechanics: Fundamentals and Applications*, McGraw Hill, Columbus, OH, 2006.

[8] Szmerekovsky, A. G., and Palazotto, A. N., "Structural Dynamics Considerations for a Hydrocode Analysis of Hypervelocity Test Sled Impacts," *AIAA Journal*, Vol. 44, No. 6, June 2006, pp. 1350–1359.

[9] Szmerekovsky, A. G., Palazotto, A. N., and Baker, W. P., "Scaling Numerical Models for Hypervelocity Test Sled Slipper-Rail Impacts," *International Journal of Impact Engineering*, Vol. 32, No. 6, 2006, pp. 928–946.

[10] Szmerekovsky, A. G., Palazotto, A. N., and Ernst, M. R., "Numerical Analysis for a Study of the Mitigation of Hypervelocity Gouging," AIAA Paper 2004-1922, April 2004.

[11] Szmerekovsky, A. G., Palazotto, A. N., and Ernst, M. R., "Numerical Analysis for a Study of the Mitigation of Hypervelocity Gouging," *Dayton-Cincinnati Aerospace Science Symposium*; modified and published as AIAA Paper 2004-1922, March 2004.

[12] Szmerekovsky, A. G., Palazotto, A. N., and Cinnamon, J. D., "An Improved Study of Temperature Changes During Hypervelocity Sliding High Energy Impact," AIAA Paper 2006-2090, May 2006.

[13] Cinnamon, J. D., and Palazotto, A. N., "Further Validation of a General Approximation for Impact Penetration Depth Considering Hypervelocity Gouging Data," *International Journal of Impact Engineering*, Vol. 34, No. 8, 2007, pp. 1307–1326.

[14] Cinnamon, J. D., Jones, S. E., House, J. W., and Wilson, L. L., "A One-Dimensional Analysis of Rod Penetration," *International Journal of Impact Engineering*, Vol. 12, No. 2, 1992, pp. 145–166.

[15] Cinnamon, J. D., Jones, S. E., Wilson, L. L., and House, J. W., "Further Results on the One-Dimensional Analysis of Rod Penetration," *Proceedings of the Sixteenth Southeastern Conference on Theoretical and Applied Mechanics, Vol. XVI: Developments in Theoretical and Applied Mechanics*, Antar, B. N. (ed), Univ. of Tennesee Space Institute, Tullahoma, TN, 1992, pp. II.11.36–II.11.43.

[16] Cinnamon, J. D., "Further One-Dimensional Analysis of Long-Rod Penetration of Semi-Infinite Targets," Master's Thesis, Dept. of Aerospace Engineering, Univ. of Texas at Austin, TX, July 1992.

[17] Cinnamon, J. D., "Further One-Dimensional Analysis of Long-Rod Penetration of Semi-Infinite Targets," Air Force Inst. of Technology, Technical Rept. AFIT/CI/CIA-92-086, Wright-Patterson AFB, Dayton, OH, 1992.

[18] Cinnamon, J. D., and Jones, S. E., "An Analysis of Rod Penetration of Semi-Infinite Targets Using an Average Pressure Estimate," *Proceedings of the ASME Pressure Vessel and Piping Conference*, Vol. 394, ASME, New York, 1999, pp. 19–35.

[19] Jones, S. E., Gillis, P. P., and Foster, J. C., "On the Penetration of Semi-Infinite Targets by Long Rods," *Journal of the Mechanics and Physics of Solids*, Vol. 35, No. 121, 1987, pp. 121–131.

[20] Kerber, M. W., Jones, S. E., Fisher, C. A., and Wilson, L. L., "Penetration and Cratering of Semi-Infinite Targets by Long Rods," *Proceedings of the 12th International Symposium on Ballistics*, International Ballistics Committee, San Antonio, TX, 1990, pp. 111–127.

[21] Murphy, M. J., "Survey of the Influence of Velocity and Material on the Projectile Energy/Hole Volume Relationship," *Proceedings of the 10th International Symposium on Ballistics*, International Ballistics Committee, San Antonio, TX, 1987, pp. 45–53.

[22] Tate, A., "A Theory for the Deceleration of Long Rods After Impact," *Journal of the Mechanics and Physics of Solids*, Vol. 15, No. 387, 1967, 389–399.

[23] Ravid, M., Bodner, S. R., and Holoman, I., "Analytical Investigation of the Initial Stages of Impact of Rods on Metallic and Ceramic Targets at Velocities of 1 to 9 km/s," *Proceedings of the 12th International Symposium on Ballistics*, International Ballistics Committee, San Antonio, TX, 1990, pp. 220–233.

[24] Nicholas, T., and Recht, R. F., "Introduction to Impact Phenomena," *High Velocity Impact Dynamics*, Wiley, New York, 1990, pp. 1–63.

[25] Zukas, J. A., Nicholas, T., Swift, H. F., Greszczuk, Longin B., and Curran, D. R., *Impact Dynamics*, Krieger, Malabar, FL, 1992.

[26] Marsh, S. P. (ed.), *LASL Shock Hugoniot Data*, Univ. of California Press, Berkeley, CA, 1980.

[27] Jones, S. E., Marlow, R. B., House, J. W., and Wilson, L. L., "A One-Dimensional Analysis of the Penetration of Semi-Infinite 1100-0 Aluminum Targets by Rods," *International Journal of Impact Engineering*, Vol. 14, Nos. 1–4, 1993, pp. 407–416.

[28] Barker, L. M., Trucano, T. G., and Munford, L. W., "Surface Gouging by Hypervelocity Sliding Contact," Sandia National Labs., Technical Rept. SAND87-1328, Albuquerque, NM, Sept. 1987.

[29] Barker, L. M., Trucano, T. G., and Susoeff, A. R., "Railgun Rail Gouging by Hypervelocity Sliding Contact," *IEEE Transactions on Magnetics*, Vol. 25, No. 1, 1988, pp. 83–87.

[30] Graff, K. F., and Dettloff, B. B., "The Gouging Phenomenon Between Metal Surfaces at Very High Speeds," *Wear*, Vol. 14, Issue 2, 1969, pp. 87–97.

[31] Krupovage, D. J., and Rasmussen, H. J., "Hypersonic Rocket Sled Development," High Speed Test Track Facility, 6585th Test Group, Technical Rept. JON: 99930000, Holloman AFB, NM, Sept. 1982.

[32] Mixon, L. C., "Assessment of Rocket Sled Slipper Wear/Gouging Phenomena," Applied Research Associates, Inc., F08635-97-C-0041, Albuquerque, NM, 1997.

[33] Tarcza, K. R., and Weldon, W. F., "Metal Gouging at Law Relative Sliding Velocities," *Wear*, Vol. 209, Issues 1–2, 1997, pp. 21–30.

[34] Tarcza, K. R., "The Gouging Phenomenon at Low Relative Sliding Velocities," Master's Thesis, Department of Mechanical Engineering, Univ. of Texas at Austin, TX, Dec. 1995.

[35] Hooser, M., "Simulation of a 10,000 Foot Per Second Ground Vehicle," AIAA Paper 2000-2290, June 2000.

[36] Cinnamon, J. D., and Palazotto, A. N., "Metallographic Examination and Validation of Thermal Effects in Hypervelocity Gouging," *ASME Journal of Pressure Vessel Technology*, Vol. 129, No. 1, 2007, pp. 133–141.

Chapter 8

Validation of Constitutive Models for High Strain Rates in Hypervelocity Impact

I. Introduction

IN CHAPTER 6, the new material flow models of VascoMax 300 and 1080 steel (based on experiment) were validated in simulations of the split Hopkinson bar (SHB) tests and the Taylor impact tests. These were midrange strain-rate events. With the successful creation of a laboratory hypervelocity gouging test, the validation of the material constitutive models can be extended into the high-strain-rate regime. Specifically, the CTH hydrocode can be used to simulate the hypervelocity gouging experiments to validate its ability, with the new flow models, to predict the observed behavior.

Prior to using CTH to model these laboratory hypervelocity gouging impacts, a discussion concerning modeling these events in a two-dimensional CTH environment is conducted. The gouging tests are then predicted using a CTH model. The ultimate goal of this effort is to validate CTH's ability to accurately predict hypervelocity gouging so that it can be used as an effective tool in the mitigation of gouging in the HHSTT scenario.

II. Examination of CTH Modeling

The effort to simulate these kind of hypervelocity impacts has been conducted for many years [1–4]. One of the most successful codes is the hydrocode CTH. CTH is unique in that it possesses a large body of experimental data embedded in equation-of-state (EOS) tables. Additionally, CTH was written as a shock-physics code whose primary purpose is to simulate high-energy impact events. The primary limitation of this code is that it is extremely computationally intensive. Users of CTH typically create state-of-the-art computer clusters, or utilize time on some of the nation's fastest supercomputers, in order to have the computation time stay in the reasonable range. A simple problem, implemented in full three-dimensionality, with all of the associated optional subroutines engaged (i.e., heat conduction, etc.) can take two weeks to generate 10 μs of data. Therefore, many users model these impact problems with plane-strain or two-dimensional axisymmetric models to reduce computation time to the order of tens of hours and days.

With regard to the specific impact problem at the HHSTT, previous researchers performed a comparison between simulating the shoe/rail interaction in two-dimensional plane strain vs a full three-dimensional implementation [5–8]. The conclusion that a plane-strain version of the impact event was significantly similar to the results from a three-dimensional solution in the time period considered (up to 20–30 μs) was drawn. Based on this analysis, subsequent model development was done in plane strain.

Szmerekovsky and colleagues [9–14] developed a two-dimensional plane-strain model of the shoe/rail interaction by taking the sled arrangement and distributing the mass of the sled across the four shoes. The mass is then taken to distribute evenly through the width of the shoe, and a unit slice is removed and modeled. This process is depicted in Fig. 8.1. This model was used very successfully to simulate the gouging phenomenon. Many of the features observed in the field and from experimental analysis of the gouges were generated by this plane-strain model in CTH [9–14]. Figure 8.2 illustrates a typical gouge created in CTH. Material mixing is created and represents one of the unique capabilities of this code. Based on the success of these plane-strain models, and the unreasonable computation times of three-dimensional implementations, a decision was made to model these impact events in two-dimensional plane strain.

As mentioned earlier, the major limitation of these earlier modeling efforts was the lack of specific material models in CTH for VascoMax 300 and 1080 steel. Although suitable EOS models exist, the constitutive models governing the strength of the materials below EOS pressures are not available. Therefore, an extensive effort was undertaken to develop these material strength models (see Chapter 5). To validate these material models, experimental Taylor impact tests were conducted (see Chapter 6). The focus of this chapter is to extend the validation into the hypervelocity impact regime.

Related to the midrange strain-rate validation of the strength models in CTH (using the Taylor impact tests), an evaluation of the suitability of two-dimensional simulations was performed. In Chapter 6, a two-dimensional model was able to match posttest geometry of the Taylor impact test within 5% (which involves a cylinder of material impacting a nondeforming target at velocities around 200 m/s).

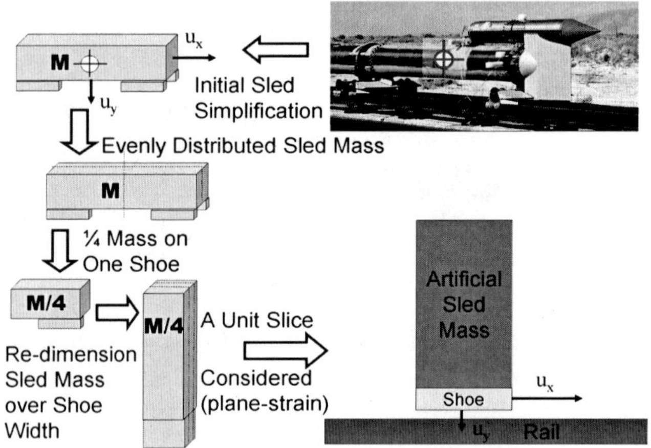

Fig. 8.1 Sled plane-strain gouging model development [13].

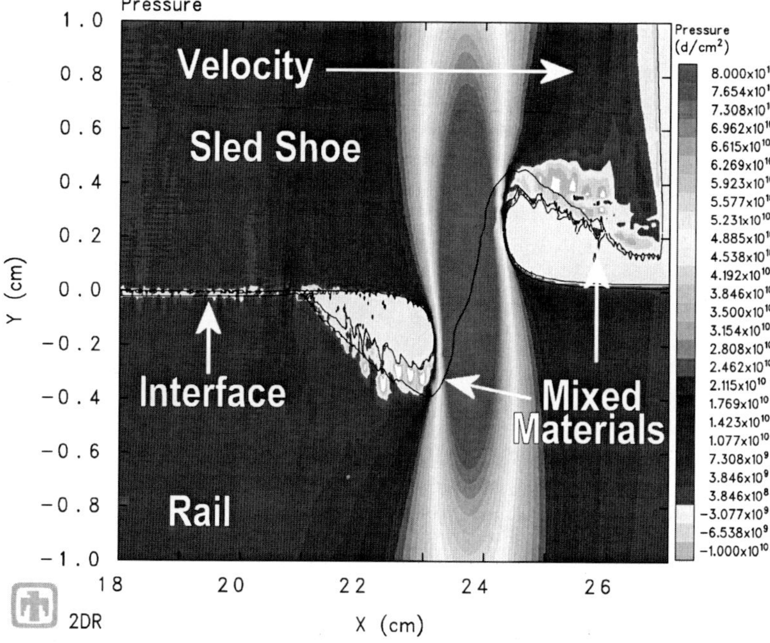

Fig. 8.2 Typical CTH gouge simulation.

Additionally, because of the nature of CTH's development, some additional features have been added to the code for specific reasons. An evaluation of the three material contact schemes was undertaken, and the most numerically stable option was identified (designated as the no-slide line option within CTH).

As already mentioned, in modeling our hypervelocity gouging tests, we are limited to choosing between plane strain and a full three-dimensional model. CTH does have a two-dimensional axisymmetric mode, but it is limited to a vertically oriented impact because it established a vertical line of symmetry and rotates a user-defined two-dimensional slice about it. Motivated by previous success in two-dimensional modeling, an investigation into the difference between an axisymmetric mode and a plane-strain mode solution to impact problems was conducted [15–17].

A. Model Mode Comparison

For this examination, a penetration model was created based on our hypervelocity experimentation. The projectiles used in our gun tests were modeled to impact the rail at a normal incidence angle and at the velocity range corresponding to the vertical impact velocity of our hypervelocity shots (i.e., 375–555 m/s) from Chapter 7. The material flow relation used in our modeling was the Zerilli–Armstrong model. The material model constants were formulated from flyer plate experiments presented in Chapter 5. Table 5.5 summarizes the material model constants used within CTH. The temperature constants c_3 and c_4 are given in electron volts, which is the unit used in CTH.

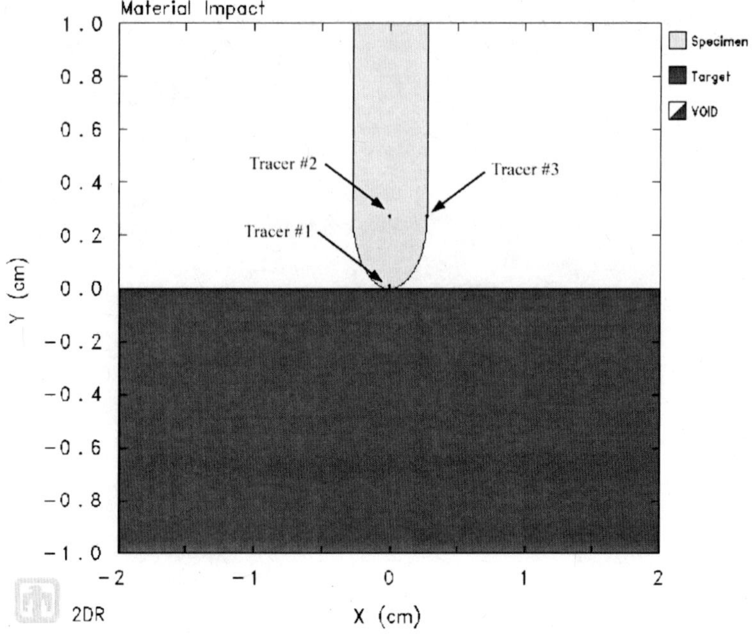

Fig. 8.3 Typical CTH solution mode comparison model.

An axisymmetric model, which represents the actual geometry of the experimental projectile, was compared against a plane-strain model of the same vertical impact. The impact geometry is depicted in Fig. 8.3. The mesh size used was 0.002 cm, which was arrived at by a previous mesh convergence study. In this work, a representative case is examined, in which the projectile hits the target at 375 m/s. Figures 8.4 and 8.5 compare the axisymmetric solution to the plane-strain

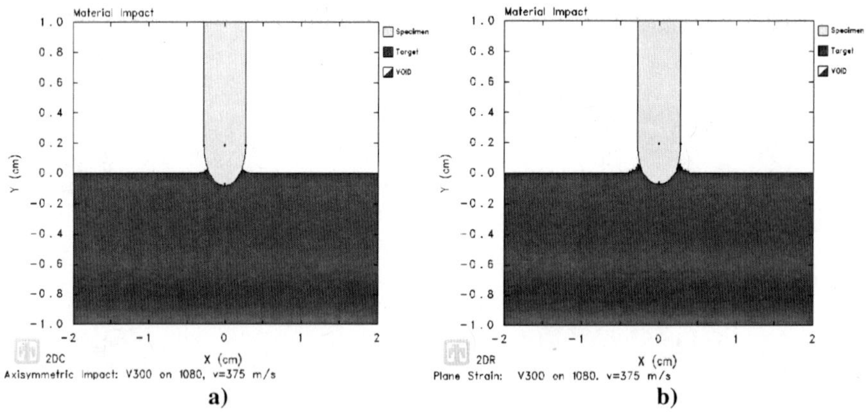

Fig. 8.4 CTH solution mode comparison model at 2.5 μs, a) axisymmetric and b) plane strain.

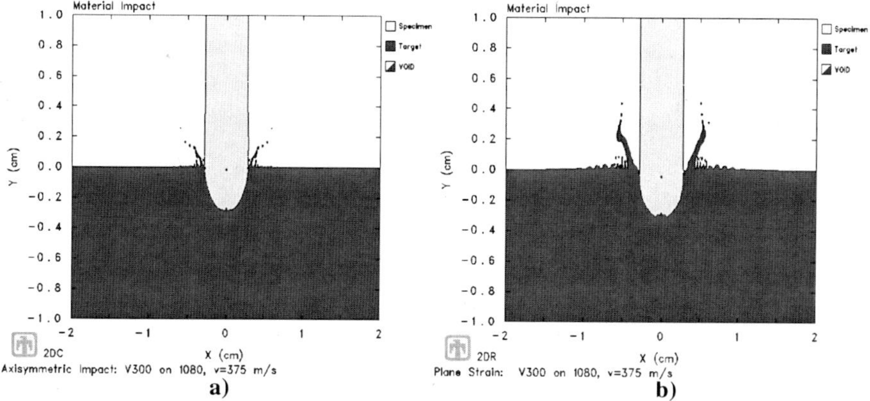

Fig. 8.5 CTH solution mode comparison model at 10 μs, a) axisymmetric and b) plane strain.

solution at discrete times (in which the axisymmetric solutions appears on the left and the plane strain on the right). Note that the penetrator deformation and penetration depth are in good agreement out to 10 μs, which corresponds both to the event time of the hypervelocity gouging impact test and the gouge event in the HHSTT scenario.

In examining the difference in cell pressure generated in each of the solution techniques, six Lagrangian grid points were selected, and the pressure was recorded at discrete points of time. Figures 8.6 and 8.7 illustrate the pressure profiles, where the black points in Fig. 8.7 indicate the points where the pressure was compared. Although there appears to be some general differences in appearance, the magnitude of the pressure variance is not excessive. Table 8.1 records the pressure differences at each point and summarizes the results.

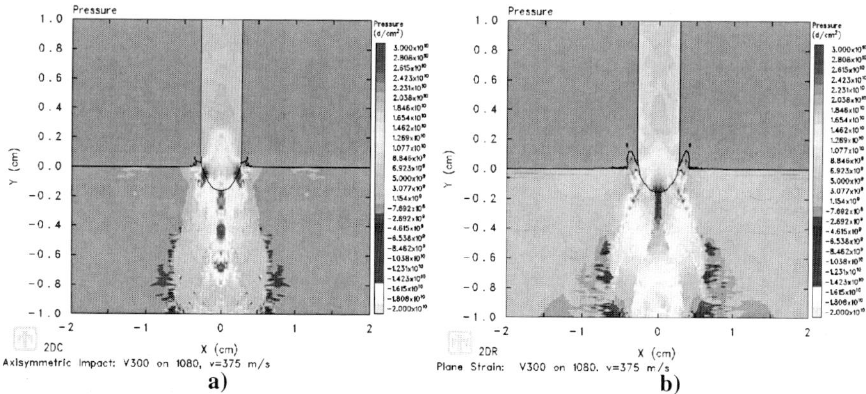

Fig. 8.6 CTH pressure solution mode comparison model at 2.5 μs, a) axisymmetric and b) plane strain.

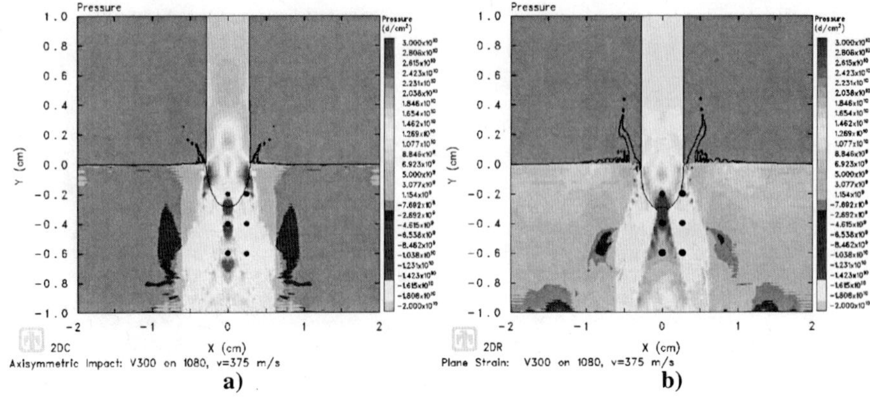

Fig. 8.7 CTH pressure solution mode comparison model at 10 μs, a) axisymmetric and b) plane strain.

So, although there is some measurable difference between the solution techniques, the magnitude of the difference is not significant. Additionally, the gross deformation predicted by the methods matches very well. Note that in Fig. 8.7 the penetrators are almost identical in diameter (0.26 cm for axisymmetric, 0.25 cm for plane strain, for a difference of 3.8%). Additionally, the penetration depth for both cases is also comparable (0.29 cm for axisymmetric, 0.3 cm for plane strain, for a difference of 3.4%). Even better agreement was noted in the 555-m/s impact case. Based on this evaluation, it can be concluded that the plane-strain solution technique can fairly accurately model this three-dimensional impact event. The two-dimensional axisymmetric model would be employed for the modeling of our laboratory hypervelocity gouging test, but our impact geometry is not compatible with the implementation in CTH.

B. Model Mode Comparison Summary

Therefore, based on this analysis, the laboratory hypervelocity impact test is modeled in two-dimensional plane strain. This approach matches those of

Table 8.1 CTH pressure solution mode comparison—plane strain compared against axisymmetric

Discrete point, x coordinate	Discrete point, y coordinate, cm	% Difference, $t = 5\,\mu s$	% Difference, $t = 10\,\mu s$
0	−0.2	21	16
0	−0.4	9	14
0	−0.6	0	16
0.25 cm	−0.2	17	30
0.25 cm	−0.4	47	0
0.25 cm	−0.6	36	12
Average:		31	14

previous investigators and has been shown to be sufficiently accurate in the preceding section. Additionally, this approach will be used to model the sled/rail gouging phenomenon for the HHSTT problem. Of course, the suitability of this approach can be judged based on its ability to accurately generate the experimental characteristics of the gouging tests.

III. Validation of CTH Hypervelocity Gouging Model

To establish the validity of CTH to model the full sled/rail interaction at the HHSTT, the newly developed material flow models and CTH are validated against the hypervelocity gouging tests presented in the Chapter 7.

A. CTH Model of Laboratory Hypervelocity Gouging Test

With the investigation just presented, a two-dimensional plane-strain model of the hypervelocity gouging test was constructed. Using this approach, the cylinder with a hemispherical nose is modeled as a unit-thickness, plane-strain plate, with a rounded leading edge. Figure 8.8 illustrates this model. The high-speed photography available from the impacts indicated that the projectile oriented during flight in such a manner as to impact the target rail as depicted in Fig. 8.8. This is illustrated in Fig. 8.9. This is a top-down view of the impact and represents the available photographic depiction. While the velocity vector was still oriented at 10

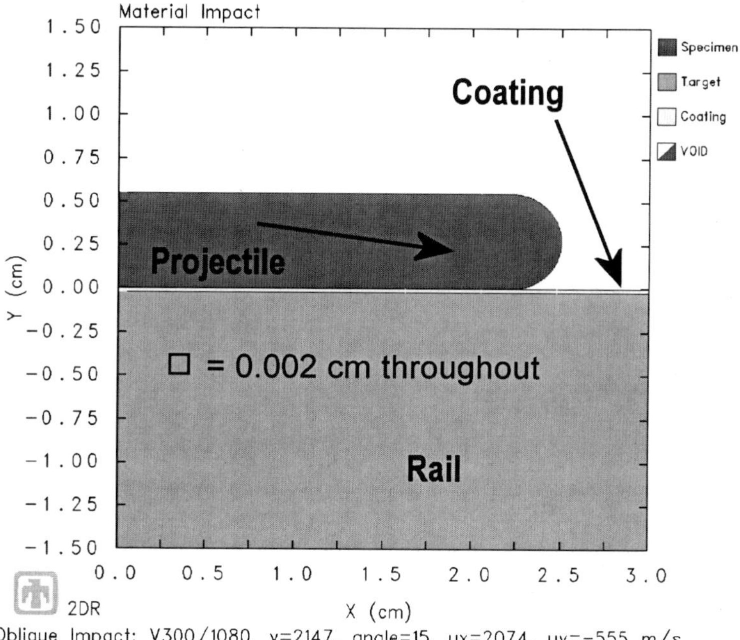

Fig. 8.8 CTH model of hypervelocity gouging test (velocity vector indicated).

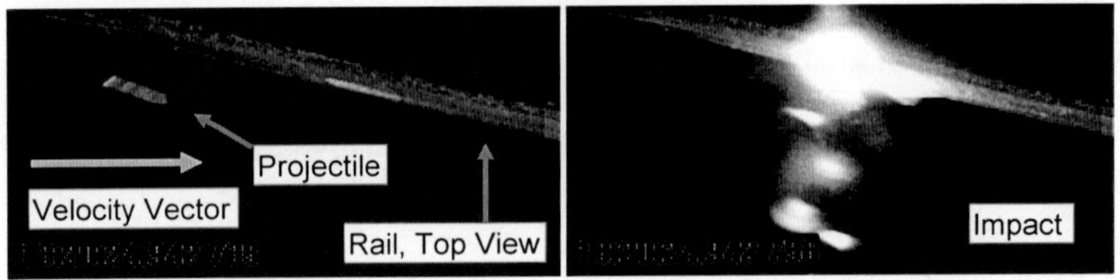

Fig. 8.9 High-speed photograph of projectile impact orientation.

or 15 deg to the rail surface, the long axis of the projectile was aligned with the longitudinal axis of the target rail.

Again, the appropriate mesh size was utilized—0.002 cm. All four impact cases from Chapter 7 were examined. A sample CTH input file is included in Appendix A.

B. CTH Simulation of Laboratory Hypervelocity Gouging Test

The effort to model the hypervelocity gouging tests was extremely successful. A representative case is presented in this section. Figure 8.10 is of impact test he-1, with an impact velocity of 2147 m/s and an impact angle of 15 deg. The solution shows the characteristic material mixing and high plasticity of gouging. Additionally, Fig. 8.11 illustrates the pressure and strain rate associated with 10 μs. Note that the strain rates computed by CTH are predominately in the 10^4/s–10^6/s range in the gouging region, but that the midrange strain rates are also indicated within the solution.

The gouge depth can be seen in Fig. 8.10 to be approximately 1.1 mm, which agrees closely with the experiments presented in Chapter 7. Indeed, the simulated results are within the uncertainty range of the measurements of the experimental gouge depths. All four simulations of the laboratory gouging shots demonstrated excellent agreement to experimentally observed values for gouge depth. Table 8.2

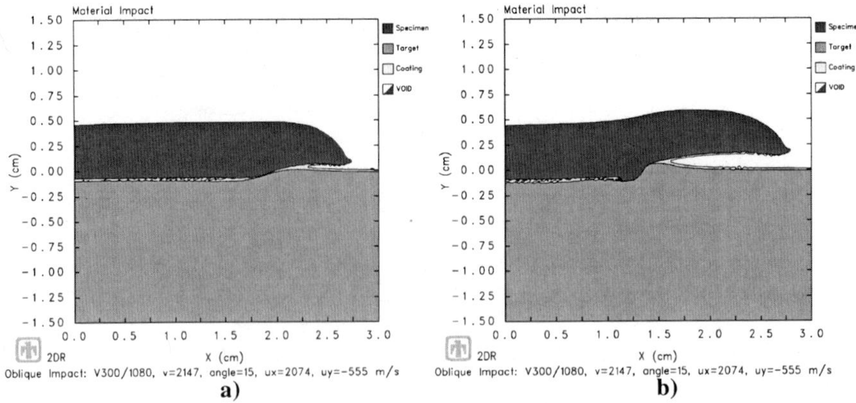

Fig. 8.10 Simulation of test he-1 at a) 5 μs and b) 10 μs.

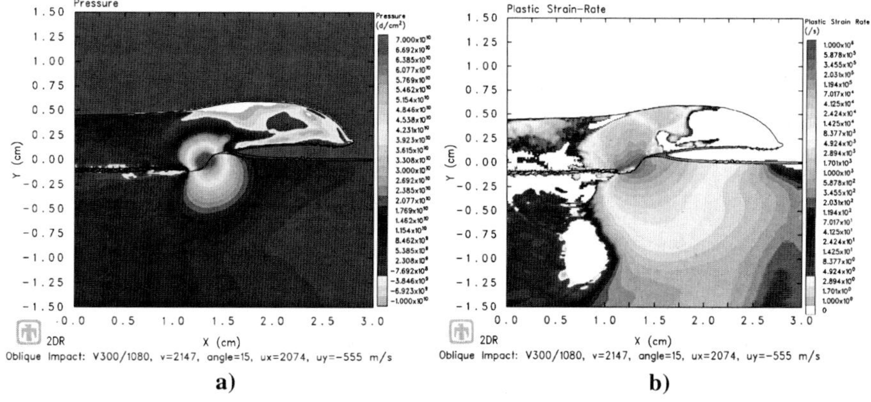

Fig. 8.11 Simulation of test he-1, a) pressure and b) strain rate at 10 μs.

summarizes these results. The one-dimensional theory predictions (using the approach outlined earlier in this chapter) are presented in this table to show the agreement between the one-dimensional analytical approach and the computational simulation. Figure 8.12 summarizes the information in graphical form.

The characteristics of hypervelocity gouging, such as a high-pressure concentration [9–13] at the gouge location, material jetting, and material mixing, are evident in these simulations. Additionally, sufficient temperature by the plasticity that could create the microstructural changes reported in Chapter 7 was generated. Figure 8.13 illustrates the temperature profile of test he-1 as generated by the CTH simulation.

C. CTH Simulation of Thermal Characteristics of the Laboratory Hypervelocity Gouging Test

As part of the CTH model validation for the laboratory hypervelocity gouging test, we can closely examine CTH's ability to match the microstructural observations from the gouges. As presented in Chapter 7, the gouges created in the scaled hypervelocity test exhibited the same kind of microstructural changes within the rail as the full-scale HHSTT gouge. Returning to these metallurgical

Table 8.2 CTH simulation of VascoMax 300 on 1080 steel test series

Test	Coating	Angle, deg	Impact velocity, m/s	Gouge depth (predicted), mm	Gouge depth (measured), mm	Gouge depth (simulated), mm
hi-1	Iron ox.	10	2225	0.51	0.5 ± 0.1	0.6
hi-2	Iron ox.	15	2150	1.03	1.0 ± 0.1	1.1
he-1	Epoxy	15	2147	1.03	1.0 ± 0.1	1.1
he-2	Epoxy	10	2163	0.48	0.5 ± 0.1	0.5

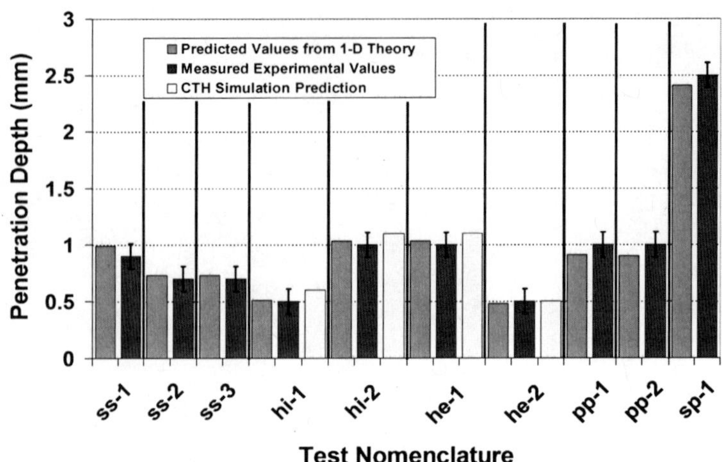

Fig. 8.12 Hypervelocity gouge test series results summary.

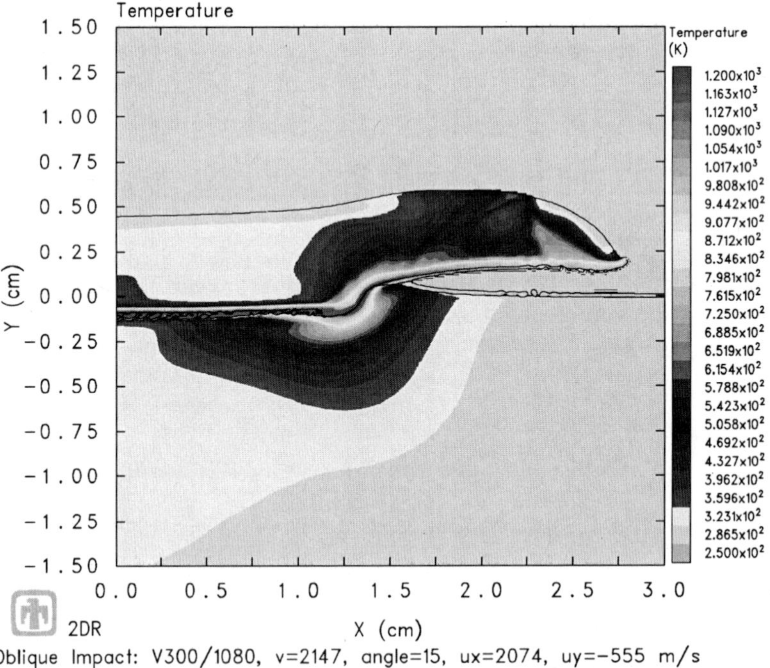

Fig. 8.13 Simulation of test he-1, temperature, at 10 μs.

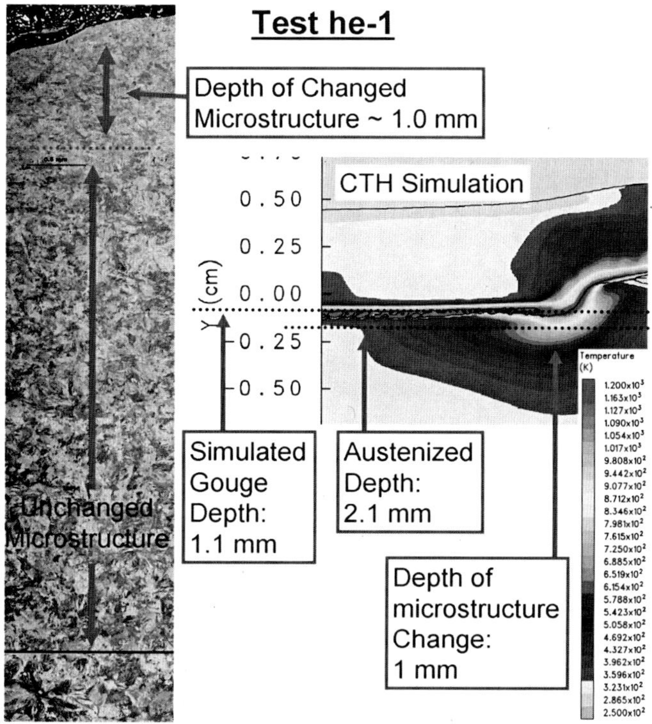

Fig. 8.14 Comparison of CTH simulation to observed microstructure, test he-1.

results, we can compare the CTH simulations of generated temperature to those microstructural changes observed.

Figure 8.14 is a summary of the comparison between the CTH simulation of test he-1 and the metallurgical results. Note that the depth of microstructure change is matched, as well as the gouge depth (which was noted in Table 8.2). The austenzing depth is discussed in Chapter 4 and occurs when the temperature exceeds 1000 K. The CTH simulation portion of that figure comes from zooming in on the gouge in Fig. 8.13.

Similar results are seen from examining tests hi-1 (in Fig. 8.15) and hi-2 (in Fig. 8.16). Again, the austenizing temperatures are matched against micrographs of the gouged rails.

D. Further Results from the Comparison of CTH Simulations to the One-Dimensional Penetration Theory

In the investigation of the hypervelocity gouging impact simulation within CTH, some additional impact conditions were examined. As noted in Chapter 7, the one-dimensional penetration theory was primary developed to estimate the laboratory configuration necessary to ensure gouge creation. To further validate the

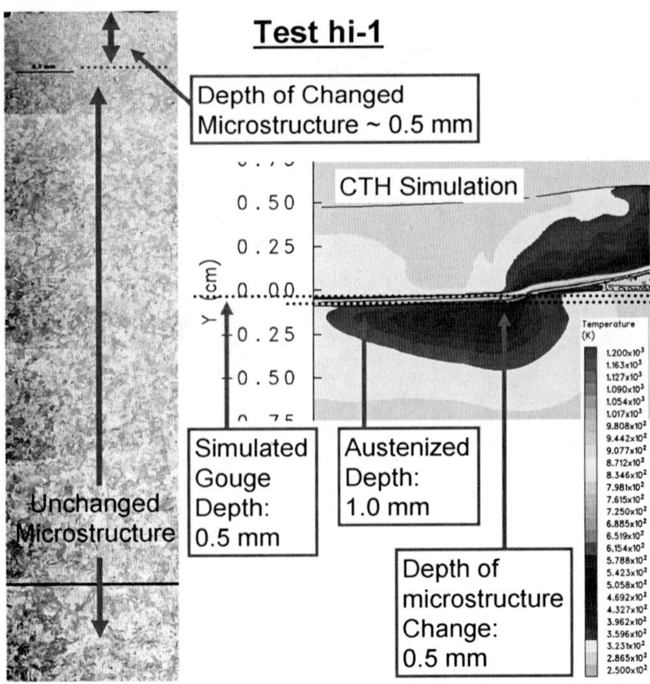

Fig. 8.15 Comparison of CTH simulation to observed microstructure, test hi-1.

one-dimensional approach, the limiting case of a 5-deg impact was simulated in CTH, as well as one of 20-deg impact.

Figure 8.17 illustrates the impact velocity of test hi-1 (2225 m/s) oriented at a 5-deg striking angle and simulated in CTH. The one-dimensional theory from Chapter 7 predicts a gouge depth of 0.044 mm, or essentially no penetration. This type of impact is predicted by CTH, in which there is some negligible deformation of the rail and associated heating from small amounts of plasticity. Note that even with very little surface deformation, the temperatures have risen above the austenizing limit, which might point to the cause of the type of microstructure changes observed in nondamaged (in-service) rails from the HHSTT seen in Chapter 4.

When the impact angle is increased to 20 deg, the resulting prediction for the one-dimensional penetration theory becomes 1.52 mm. Figure 8.18 depicts how CTH creates the same result. In this case, more plasticity and gouge depth are generated, and the temperature profile is correspondingly more significant.

The fidelity of the one-dimensional penetration theory has been demonstrated to be beyond the several cases in which experimental tests were conducted. That is, the theory can be applied to a broader range of impact conditions and shows promise in the study of impacts that do not generate gouging but create "wear."

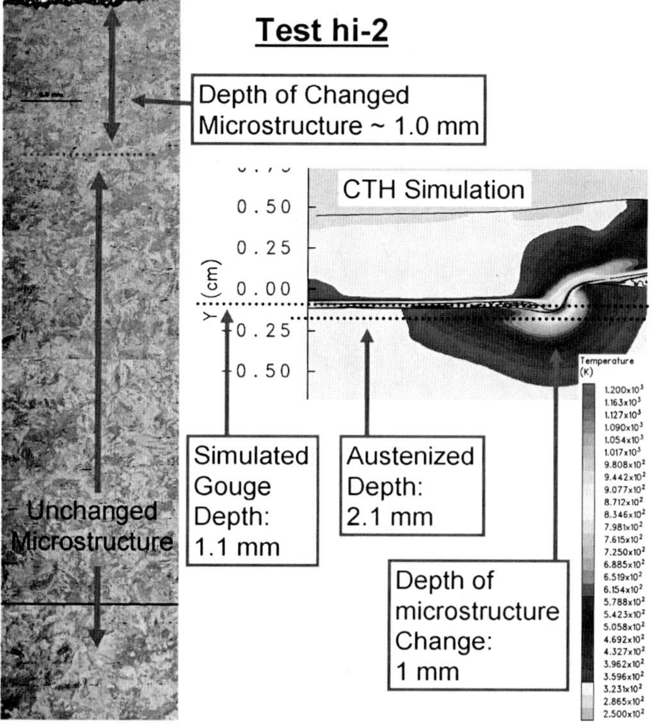

Fig. 8.16 Comparison of CTH simulation to observed microstructure, test hi-2.

E. Summary of Validation of CTH Hypervelocity Gouging Model

A CTH model was validated against a series of laboratory hypervelocity gouging experiments. The simulations were able to match both gouge depth and temperature profiles generated by the plastic deformation of the materials. Additionally, the one-dimensional penetration theory was shown to successfully predict impacts beyond the range of impact conditions used in the experimental tests.

IV. Validation of Constitutive Models and CTH for Hypervelocity Modeling

The material constitutive models for VascoMax 300 and 1080 steel were tested in the high-strain-rate regime using hypervelocity gouging impact experiments. The techniques and justification for modeling these impact events in a two-dimensional mode within CTH were presented. Using the full-range Zerilli–Armstrong flow models developed in this study, CTH demonstrated its capability to accurately capture the development of gouging, the resulting gouge depth (rail damage), and the temperature inputs that resulted in observed alteration to the target's microstructure.

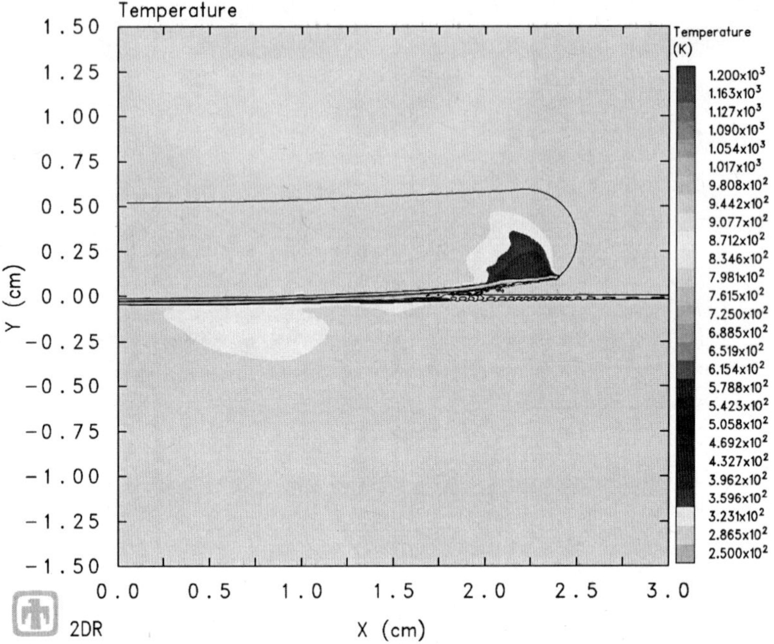

Fig. 8.17 Simulation of test he-1, 5-deg striking angle, temperature, at 10 μs.

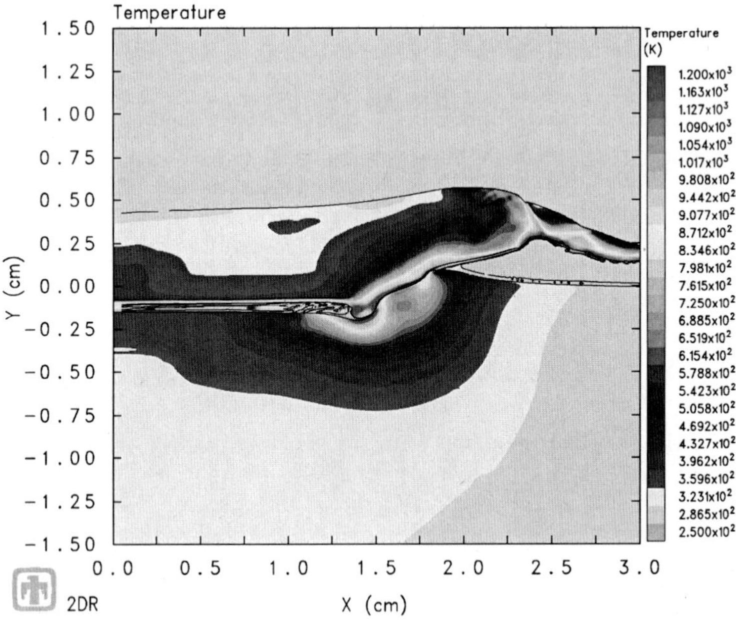

Fig. 8.18 Simulation of test he-1, 20-deg striking angle, temperature, at 10 μs.

Therefore, CTH and the constitutive models have been validated using experimentation for use in modeling hypervelocity gouging impacts. With this capability, the HHSTT sled scenario can be confidently modeled and evaluated.

References

[1] Hertel, E. S., Bell, R. L., Elrick, M. G., Farnsworth, A. V., Kerley, G. I., McGlaun, J. M., Petney, S. V., Silling, S. A., Taylor, P. A., and Yarnington, L., "CTH: A Software Family for Multidimensional Shock Physics Analysis," *Proceedings of the 19th International Symposium on Shock Waves*.

[2] McGlaun, J. M., Thompson, S. L., and Elrick, M. G., "CTH: A Three-Dimensional Shock Wave Physics Code," *International Journal of Impact Engineering*, Vol. 10, 1990, pp. 351–360.

[3] Silling, S. A., "Stability and Accuracy of Differencing Schemes for Viscoplastic Models in Wavecodes," Sandia National Labs., Technical Rept. SAND91-0141, Albuquerque, NM, 1991.

[4] Silling, S. A., "An Algorithm for Eulerian Simulation of Penetration," *New Methods in Transient Analysis*, PVP-Vol. 246, AMD-Vol. 143, ASME, Anaheim, CA, June 1992, pp. 123–128.

[5] Laird, D., and Palazotto, A. N., "Effects of Temperature on the Process of Hypervelocity Gouging," *AIAA Journal*, Vol. 41, No. 11, 2003, pp. 2251–2260.

[6] Laird, D., and Palazotto, A. N., "Gouge Development During Hypervelocity Sliding Impact," *International Journal of Impact Engineering*, Vol. 30, Issue 2, Feb. 2004, pp. 205–223.

[7] Laird, D. J., "The Investigation of Hypervelocity Gouging AFIT/DS/ENY 02-01," Ph.D. Dissertation, Department of Aeronautics and Astronautics, Air Force Inst. of Technology, Wright-Patterson AFB, Dayton, OH, March 2002.

[8] Laird, D. J., and Palazotto, A. N., "Temperature Effects on the Gouging and Mixing of Solid Metals During Hypervelocity Sliding Impact," AIAA Paper 2002-1691, April 2002.

[9] Szmerekovsky, A. G., and Palazotto, A. N., "Structural Dynamics Considerations for a Hydrocode Analysis of Hypervelocity Test Sled Impacts," *AIAA Journal*, Vol. 44, No. 6, June 2006, pp. 1350–1359.

[10] Szmerekovsky, A. G., Palazotto, A. N., and Baker, W. P., "Scaling Numerical Models for Hypervelocity Test Sled Slipper-Rail Impacts," *International Journal of Impact Engineering*, Vol. 32, No. 6, 2006, pp. 928–946.

[11] Szmerekovsky, A. G., Palazotto, A. N., and Ernst, M. R., "Numerical Analysis for a Study of the Mitigation of Hypervelocity Gouging," AIAA Paper 2004-1922, April 2004.

[12] Szmerekovsky, A. G., Palazotto, A. N., and Ernst, M. R., "Numerical Analysis for a Study of the Mitigation of Hypervelocity Gouging," *Dayton-Cincinnati Aerospace Science Symposium*, modified and published as AIAA Paper 2004-1922, March 2004.

[13] Szmerekovsky, A. G., "The Physical Understanding of the Use of Coatings to Mitigate Hypervelocity Gouging Considering Real Test Sled Dimensions AFIT/DS/ENY 04-06," Ph.D. Dissertation, Department of Aeronautics and Astronautics, Air Force Inst. of Technology, Wright-Patterson AFB, Dayton, OH, Sept. 2004.

[14] Szmerekovsky, A. G., Palazotto, A. N., and Cinnamon, J. D., "An Improved Study of Temperature Changes During Hypervelocity Sliding High Energy Impact," AIAA Paper 2006-2090, May 2006.

[15] Cinnamon, J. D., Palazotto, A. N., Szmerekovsky, A. G., and Pendleton, R. J., "Investigation of a Scaled Hypervelocity Gouging Model and Validation of Material Constitutive Models," *AIAA Journal*, Vol. 45, No. 5, 2007, pp. 1104–1112.

[16] Cinnamon, J. D., Palazotto, A. N., Szmerekovsky, A. G., and Pendleton, R. J., "Further Investigation of a Scaled Hypervelocity Gouging Model and Validation of Material Constitutive Models," AIAA Paper 2006-2087, May 2006.

[17] Pendleton, R. J., "Validation of a Scaled Plane Strain Hypervelocity Gouging Model AFIT/GAE/ENY/06-M26," Master's Thesis, Department of Aeronautics and Astronautics, Air Force Inst. of Technology, Wright-Patterson AFB, Dayton, OH, March 2006.

Chapter 9

Simulation of HHSTT Hypervelocity Gouging Scenario

I. Introduction

WITH the development of specific material constitutive models for VascoMax 300 and 1080 steel and the validation of the ability of CTH to accurately model midrange and high-strain-rate impacts, the full-scale HHSTT sled gouging problem can be confidently simulated in CTH. The previously developed CTH model for the shoe/rail interaction [1] was modified to include the new material flow models. Various impact cases were examined in order to replicate gouging in the CTH simulation that matches the experimental observations noted from HHSTT gouges. Based on the comparison of code predictions and the experimental record, conclusions are drawn concerning the character of the gouging experienced at the HHSTT [2, 3].

Gouge modeling efforts previously conducted had concluded that gouging develops within the CTH solution when either sufficient vertical impact velocity or rail discontinuities were used [1, 4]. This was verified with the new material flow models and was simulated as three gouging cases. The first was to create a vertical impact velocity sufficient to initiate gouging. The second was an impact with an angle of incidence to the rail. The final case was created when the shoe encountered a rail discontinuity. For each case, the results are presented, and comparisons to the experimental gouge characterization from Chapter 4 are made. Because the goal of the HHSTT is to extend the velocity of the sled program to 3 km/s, all simulations are conducted at that velocity.

II. CTH Modeling of the HHSTT Sled Scenario

Previous efforts in modeling the HHSTT sled gouging problem had not only yielded CTH as the optimum choice for modeling hypervelocity gouging, but a full-scale impact simulation was developed [1, 5–9]. This model was a two-dimensional plane-strain implementation, based on the conclusions made previously that the two-dimensional solution was substantially similar to results observed in a full three-dimensional simulation [4, 10, 11]. In Chapter 8, further favorable comparisons were made in the scaled hypervelocity impact test simulations between two-dimensional axisymmetric and two-dimensional plane strain. As with that analysis, the sled simulation will, over time, become more inaccurate as the reflected stress waves are not modeled in the two-dimensional plane-strain case.

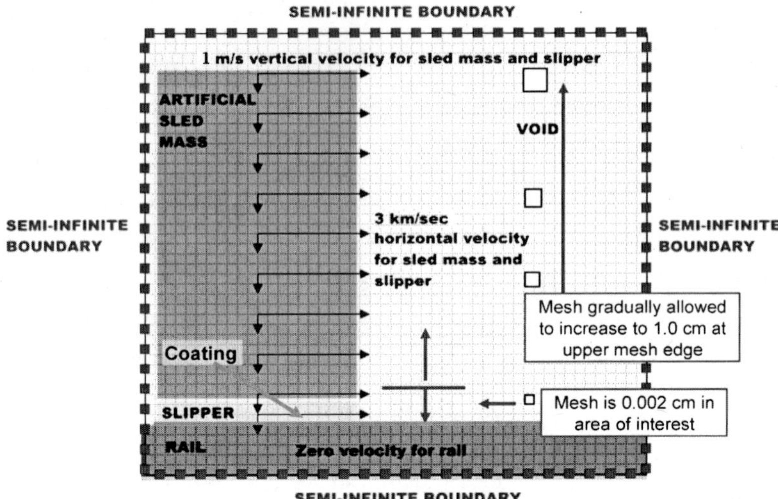

Fig. 9.1 Mesh for CTH simulation of HHSTT scenario.

Szmerekovsky computed the time for those reflected wave to return to the gouge area to be 30 μs. Therefore the gouge simulations are typically considered accurate to a time of 20 μs.

Figure 8.1, adapted from [1], illustrates how the full three-dimensional dimensionality of the sled was reduced to a plane-strain model. The plane-strain model was modeled in CTH in a mesh geometry that was the result of a convergence study conducted by Szmerekovsky. Figure 9.1 illustrates the boundary conditions and mesh used to model the sled/rail interactions [1].

The previously developed CTH model was modified to include the newly developed, strain-rate dependent Zerilli–Armstrong constitutive models for VascoMax 300 and 1080 steel. The EOS models, based on high-energy experimentation (and which include nonequilibrium thermodynamic effects), were used for VascoMax 300 and 1080 steel. The EOS for the epoxy coating was also employed, as well as a elastic perfectly plastic flow model used successfully in the past simulations [12–15]. Various impact scenarios were then investigated to better understand the gouging phenomenon. A sample CTH input file for these impact cases appears in Appendix B.

III. Gouging Case 1: Vertical Impact

In the first case, the HHSTT sled model is considered at various vertical impact velocities. In previous sled simulations [1], a vertical impact velocity of 1–2 m/s was evaluated, based on the results from an aerodynamic modeling effort [16] (Hooser, Michael, "Dynamic Design and Analysis System Simulations," unpublished data from Holloman AFB, NM, 2001). As indicated in Chapter 7, the one-dimensional penetration analysis indicates that this impact velocity would begin to deform the rail. However, the threshold kinetic energy (or threshold vertical impact velocity) for penetration is not necessarily the threshold for gouge initiation. As previous researchers have noted (see Chapter 2), the development of a hump of material in front of the shoe is critical to gouge development. As another validation of the one-dimensional penetration theory, comparisons between CTH and the theory will be presented in this section.

A. Vertical Velocity of 2 m/s

The first impact case considered was the 2-m/s vertical impact (with the 3-km/s downrange velocity). The one-dimensional theory estimated a 0.002-mm deformation—or a negligible penetration depth. Figure 9.2 illustrates the CTH simulation results. There is some wear, or localized material removal, in the coating surface, but no appreciable damage to the rail.

B. Vertical Velocity of 10 m/s

Increasing the vertical impact velocity to 10 m/s increased the deformation into the rail surface, but did not initiate gouging because the required hump of material that begins the material jetting was not created. The one-dimensional penetration analysis predicts a penetration depth of 0.06 mm for this case. Figure 9.3 illustrates that this depth was reflected by CTH as 0.1 mm and also highlights the temperature generated by the localized plasticity and nongouging deformation.

Note that the temperatures generated exceed the austenizing temperature of 1000 K.

C. Vertical Velocity of 40 m/s

Increasing the sled vertical velocity beyond approximately 40 m/s increases the damage to the rail, but remains insufficient to generate gouging. In this case, the one-dimensional model predicts a penetration depth of approximately 0.95 mm. Figure 9.4 illustrates the CTH results, showing approximately a 1-mm deformation

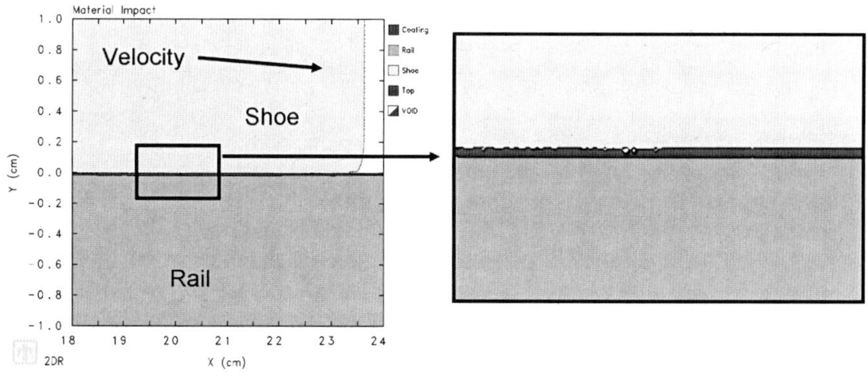

Fig. 9.2 Sled, vertical impact of 2 m/s at 10 μs.

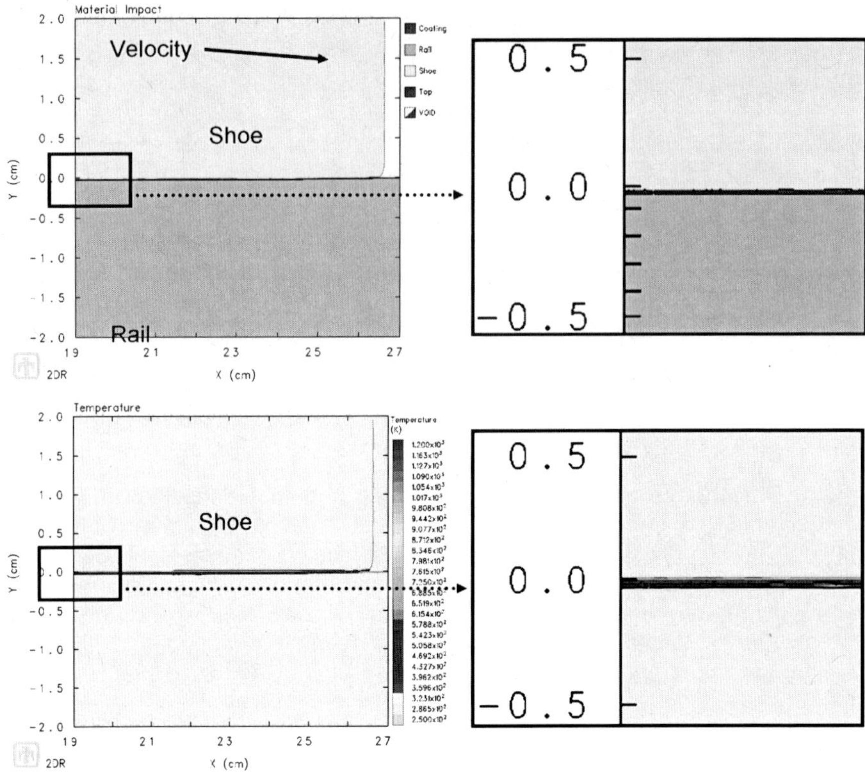

Fig. 9.3 Sled, vertical impact of 10 m/s at 20 μs.

in the rail. As with the preceding figure, the resulting temperature profile is also depicted. Note that, once again, localized temperature effects into the depth of the rail exceed 1080 steel's austenizing temperature.

D. Vertical Velocity of 75 m/s

The particular vertical velocity for gouging was not specifically ascertained in this study, but as the vertical velocity was increased to 75 m/s, gouging was observed. The one-dimensional theory predicts a penetration depth of 3.31 mm for this impact energy level. Figure 9.5 highlights the results of the simulation in terms of deformation. The penetration depth is approximately 3.5 mm in the region that is not involved in vertical material flow into the shoe and rail that characterizes gouging. The characteristics of material mixing are clear in this figure.

Similar to other investigations, it appears that gouging begins with both the creation of a material hump and the contact of the shoe and rail materials with no intervening coating layer. This is illustrated in Fig. 9.6.

The characteristic "high-pressure core" mentioned in previous work was also observed and appears in Fig. 9.7. Note that the units in CTH are dynes/cm^3;

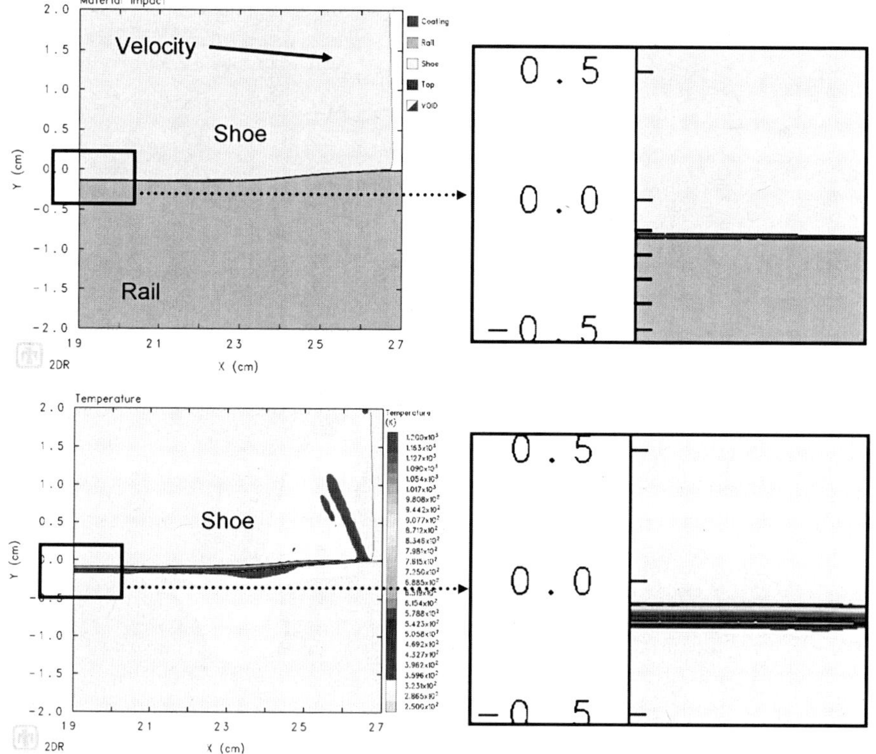

Fig. 9.4 Sled, vertical impact of 40 m/s at 20 μs.

therefore, the maximum value for this particular plot is 7 GPa when converted to more standard units.

The temperature profile that results from this plasticity is depicted in Fig. 9.8. Note that a large portion of the rail is heated above the austenizing temperature. Comparing this temperature to the experimental gouge analysis in Chapter 4, Fig. 4.5, we arrive at Fig. 9.9. This particular impact has generated results that match very closely to the metallurgical observations and subsequent thermal history that was suggested in Chapter 4.

This result is a key validation of both the material flow models and CTH's ability to model hypervelocity impact. As we will see later in this chapter, this kind of result can be obtained from the other type of impact cases. So, although a 75-m/s vertical impact might not match well with aerodynamic models of the sled, other cases will match HHSTT conditions more closely.

E. Vertical Velocity of 100 m/s

Increasing the vertical impact velocity simply increases the impact depth and the size of the resulting gouge. Figure 9.10 illustrates the resulting deformation. At this impact velocity the one-dimensional impact theory predicts a penetration depth

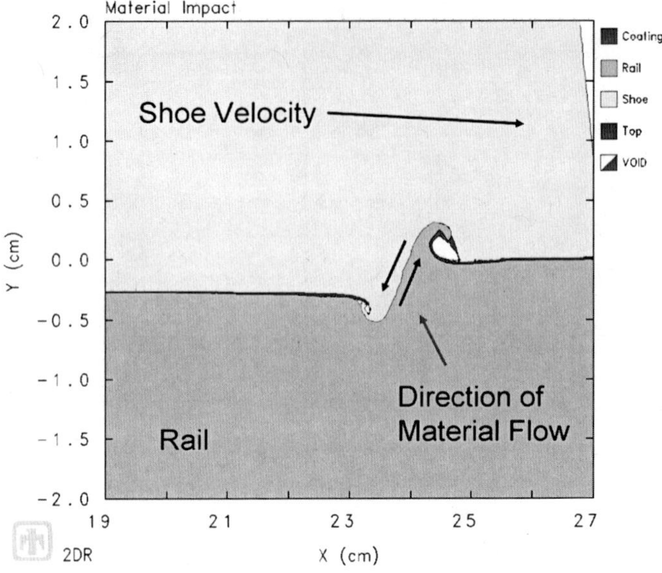

Fig. 9.5 Sled, vertical impact of 75 m/s at 20 μs.

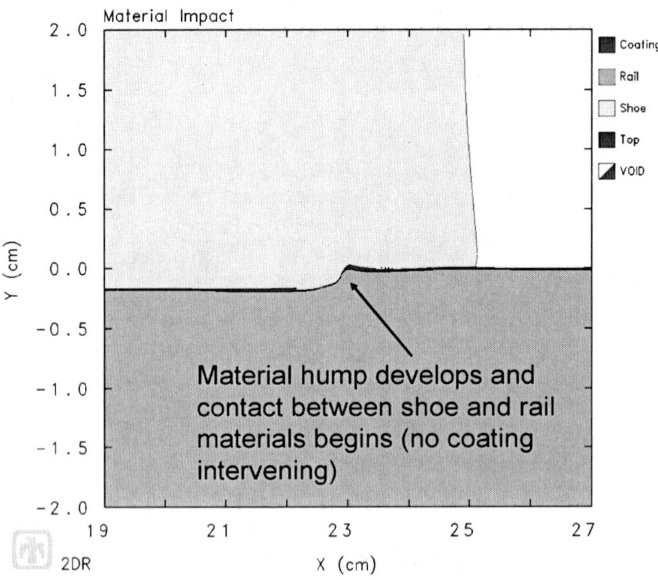

Fig. 9.6 Sled, vertical impact of 75 m/s at 14 μs.

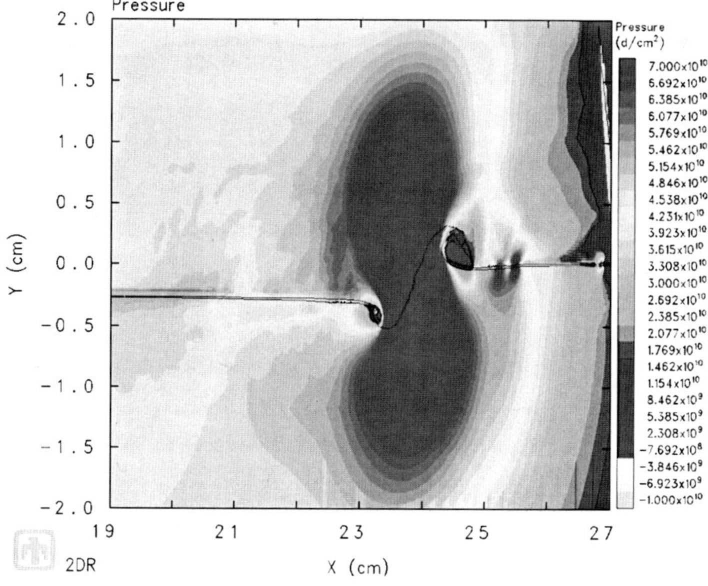

Fig. 9.7 Pressure, vertical impact of 75 m/s at 20 μs.

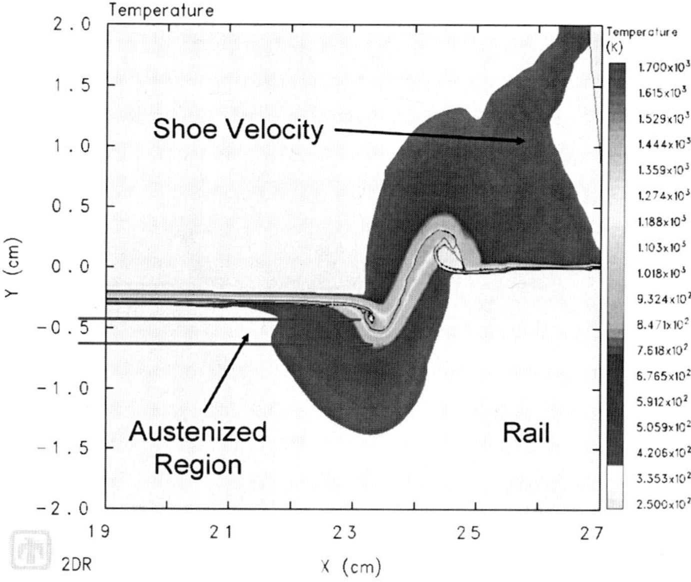

Fig. 9.8 Temperature, vertical impact of 75 m/s at 20 μs.

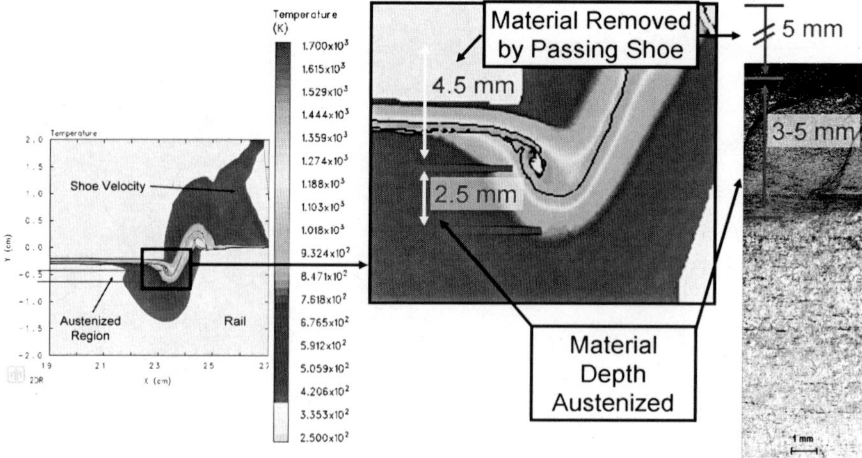

Fig. 9.9 Temperature profile comparison, vertical impact of 75 m/s at 20 μs.

of 5.8 mm. The CTH simulation indicates a penetration depth of approximately 4.5 mm. This result and those of higher impact velocities show a departure from the predicted penetration depths based on the one-dimensional theory. This is because at these impact velocities the penetration event is still occurring at 20 μs, where

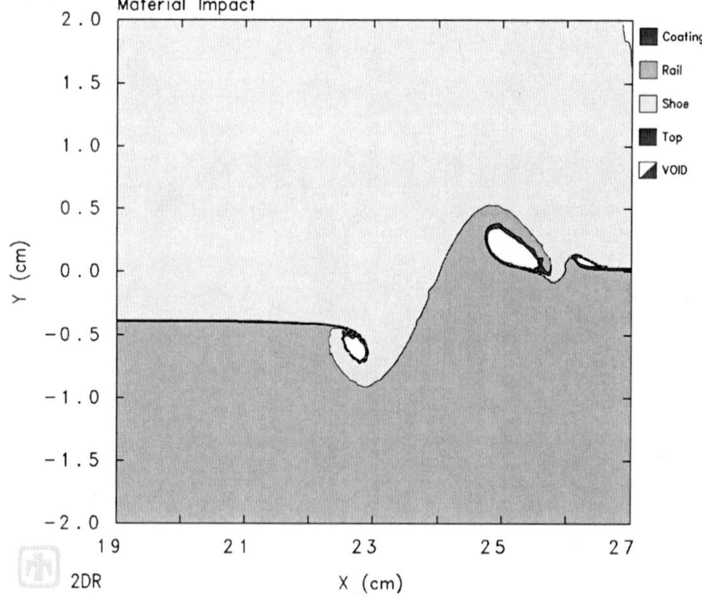

Fig. 9.10 Sled, vertical impact of 100 m/s at 20 μs.

we end the simulation. So, although the one-dimensional theory might, in fact, predict the actual resulting depths more accurately, we cannot compare against CTH beyond about 75 m/s.

Figure 9.11 shows the pressure generated by this higher energy impact, and Fig. 9.12 illustrates the resulting temperature profile.

Again, the temperature profile can be compared against the metallurgical examination of the HHSTT gouge (see Fig. 9.13). The results match fairly well for this particular case of gouging caused by high vertical impact velocity.

F. Vertical Velocity of 30 m/s on 10% Surface Area

As part of the examination of the gouging phenomenon, a smaller contact surface area was considered. The plane-strain model presented in Fig. 8.1 was redimensionalized so that the mass (200 kg) is carried by a shoe that is only 10% as long. Recalling the discussion in Chapter 7 concerning the one-dimensional penetration theory's prediction of increased penetration depth with reduced impact area, a CTH model was created to emulate that phenomenon in plane strain.

The one-dimensional penetration prediction for a 30-m/s impact on 10% of the shoe surface area is 5.32 mm. Figure 9.14 illustrates the result of the CTH simulation, which yields approximately 4.5 mm of penetration depth. Figure 9.15 illustrates the resulting pressure and temperature profiles.

Therefore, a decreased area of contact with the same mass and impact conditions will increase the penetration depth and enhance the likelihood of gouge development.

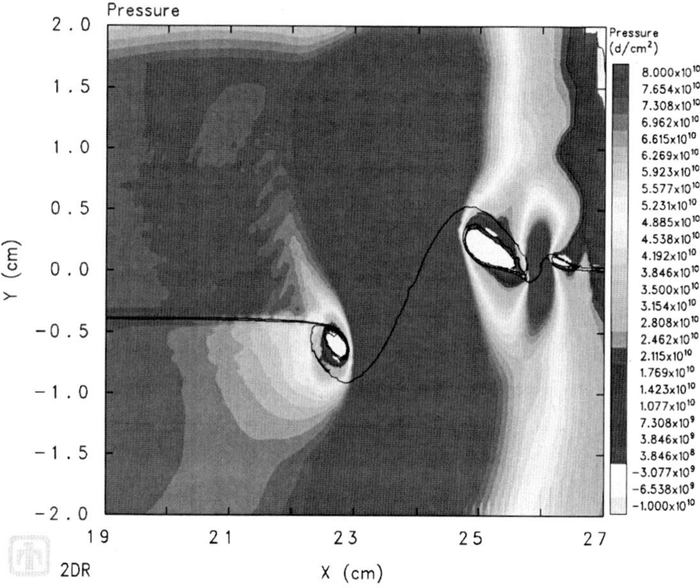

Fig. 9.11 Pressure, vertical impact of 100 m/s at 20 μs.

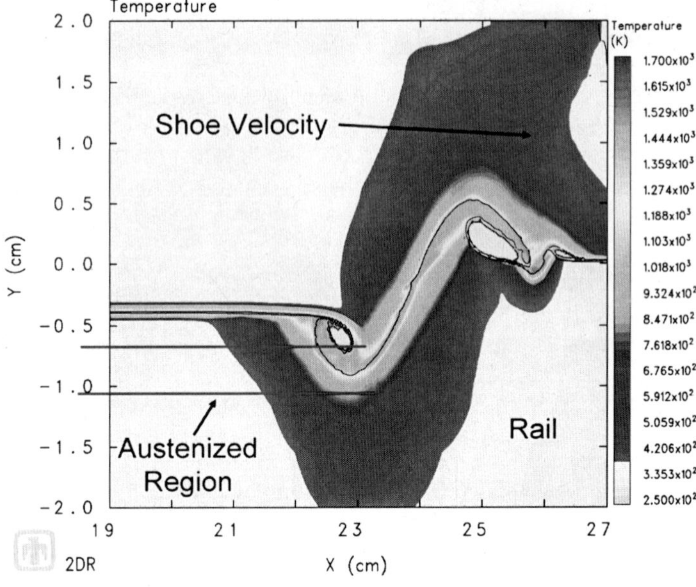

Fig. 9.12 Temperature, vertical impact of 100 m/s at 20 μs.

G. Vertical Velocity of 100 m/s with Heated Shoe

One of the areas of continuing research in this field is the exploration of the effects of having the sled shoe heat during its run. Most simulations of these hypervelocity gouges involve room-temperature materials coming into contact. For this particular evaluation, the sled shoe was heated to 1200 K within CTH

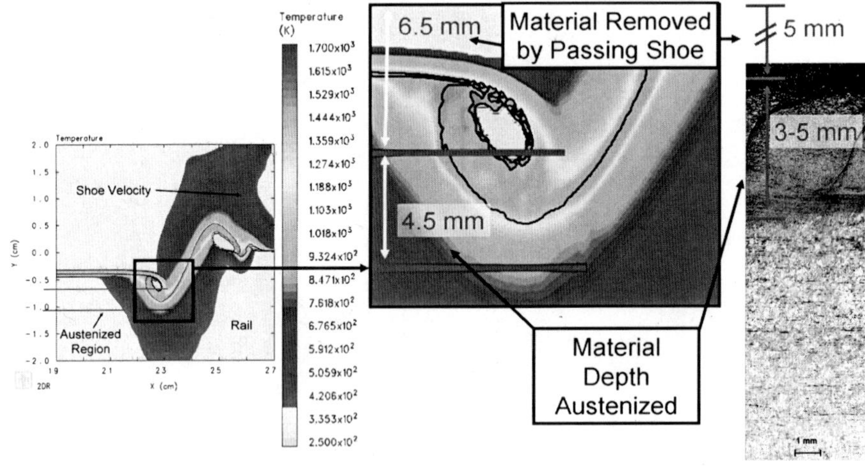

Fig. 9.13 Temperature profile comparison, vertical impact of 100 m/s at 20 μs.

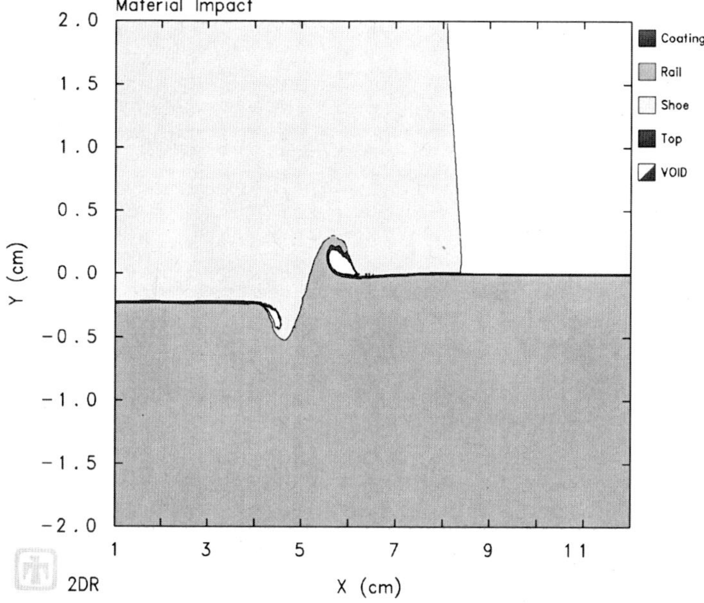

Fig. 9.14 10% Sled, vertical impact of 30 m/s at 20 μs.

prior to initiating a 100-m/s vertical impact velocity. This temperature represents a midrange value from the frictional analysis of Laird [4].

Figure 9.16 presented the results of the heated shoe run along with the preceding one in which the shoe was not heated. Although small differences can be noted (for instance, there is a slight increase in gouge depth, austenized rail depth, and gouge depth into the shoe), they are relatively small variations in the solution. Although

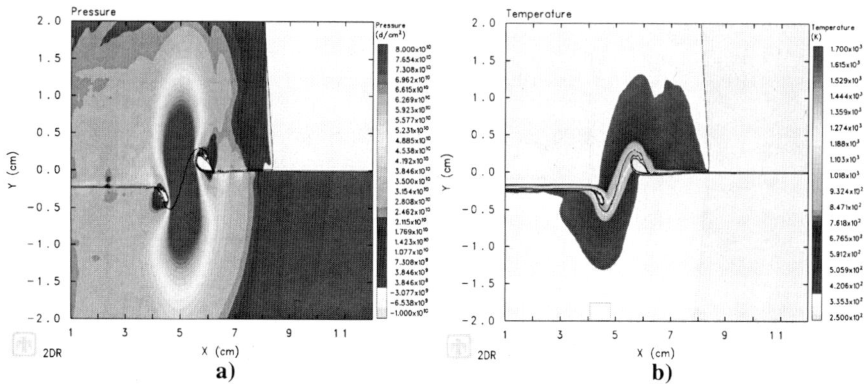

Fig. 9.15 10% Sled, vertical impact of 30 m/s at 20 μs, a) pressure and b) temperature.

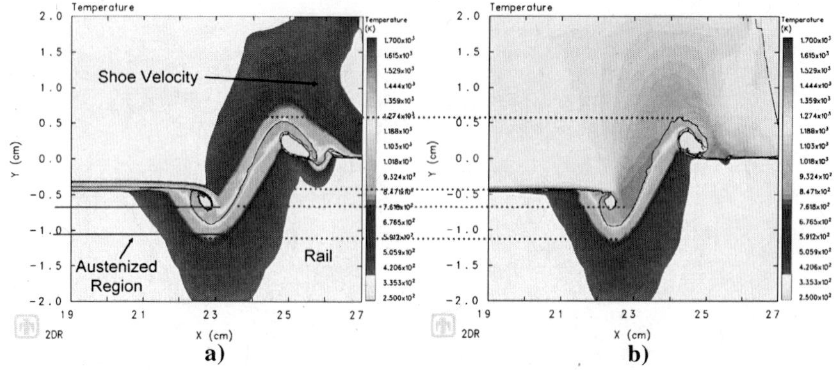

Fig. 9.16 Temperature profile comparison for heated shoe; vertical impact of 100 m/s at 20 μs, a) unheated shoe and b) heated shoe.

this is certainly not an exhaustive study on the topic, it is interesting to note the small amount of contribution the heated shoe had to the overall results.

H. Summary of Vertical Impact Case

This section was a survey of the effect of the vertical impact velocity on the initiation of gouging with the new material constitutive models. Because the one-dimensional penetration model applied to these cases, a discussion of its predictions for each case was included.

Based on the estimated vertical impact velocities present in the HHSTT scenario (from aerodynamic models [16]), the CTH simulations indicate that gouging would not occur. If the impact velocities are increased, or the contact surface area is decreased, gouging does occur. The gouging events simulated do match the character and the thermal profiles seen experimentally. In addition, until the impact velocities are increased to a level in which the vertical penetration does not finish within 20 μs, the one-dimensional penetration theory fairly accurately predicted the results. Table 9.1 summarizes the comparison.

Table 9.1 Comparison of penetration depth predictions

Vertical velocity, m/s	z (one-dimensional theory), mm	z (CTH simulation), mm
2	0.002	0.0
10	0.06	0.1
40	0.95	1.0
75	3.31	3.5
30 (10% shoe)	5.32	4.5
100	5.8	4.5

IV. Gouging Case 2: Angled Impact

A second type of impact that the sled could experience, and which does lead to gouging, is an "angled" impact. In this case, there is an angle of incidence between the rail and shoe such that the leading edge of the shoe contacts the rail first, before the remainder of the shoe impacts. The vertical velocity considered was 1 m/s.

One way in which this scenario can occur in the field is a rail height change over a length of rail. According to the HHSTT, the allowable rail height change is 0.025 in. over a length of 52 in. (0.0635 cm over a length of 132 cm) (Hooser, Michael, "HHSTT Track Tolerances," unpublished data from Holloman AFB, NM, 2006), which is an angle of incidence of 0.03 deg (if considered over the entire length). The angle would, of course, be greater if the rail height change occurred over a smaller distance. As long as the height variation occurred within the entire test distance of 132 cm and was under 0.0635 cm in magnitude, the rail would be in tolerance. Therefore, a wide range of angles is possible. This tolerance will be referred to as the "rail height tolerance" in this work and is depicted in Fig. 9.17.

Another manner in which an angled impact could occur is that the shoe gap can allow the rear of the sled to rise and thereby create an angle with the rail. This will be described in detail in Sec. IV.A.

Finally, this could also occur if a sloping rail discontinuity were encountered. According to the HHSTT, at the rail section seams, a sloping discontinuity of 0.075 in. over a 1-in. span (0.19 cm over 2.54 cm) is allowable (Hooser, Michael, "HHSTT Track Tolerances," unpublished data from Holloman AFB, NM, 2006), which is an angle of 4.29 deg. Of course, if the discontinuity were joined with a smaller angle, it would be within tolerance. This tolerance will be referred to as the "rail seam tolerance" in this work and is illustrated in Fig. 9.18. Although a brief discussion of this type of impact is presented in this section, a much more detailed examination of the rail discontinuity case is presented in the next section.

Both of these rail tolerances can be depicted in a graphical format, presented in Fig. 9.19. In this diagram, the face angle of the discontinuity is shown as a function of the discontinuity magnitude. The shaded areas represent the "operating area" where the rail discontinuity is within tolerances.

A. Incidence Angle of 0.14 deg

To compute the maximum allowable incidence angle allowed in the HHSTT scenario considering only the shoe gap scenario, a simple computation is made. Taking the maximum shoe gap of 0.635 cm (which is the maximum gap between

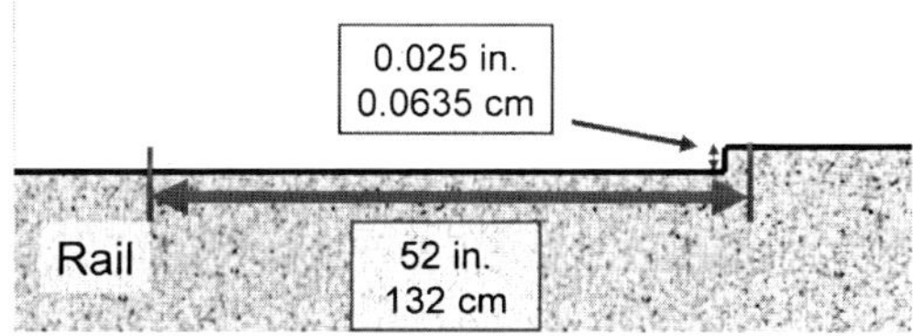

Fig. 9.17 Rail-height-tolerance illustration.

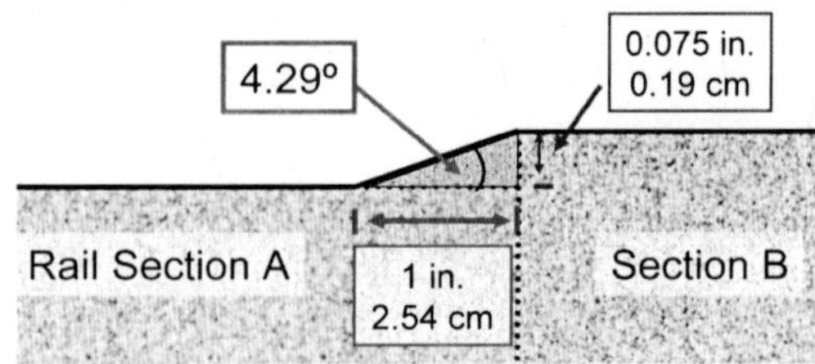

Fig. 9.18 Rail-seam-tolerance illustration.

the shoe and the rail when the shoe is full travel in the up position) and the 2.5 m between the front and back shoes on the sled (see Fig. 8.1), the maximum incidence angle is 0.14 deg. That is, if the sled is pitching down from the full-up travel position, the leading edge of the front shoe can impact the rail at 0.14 deg.

Two models of this impact were evaluated, with very similar results. One was to angle the shoe and velocity vector within CTH and have the shoe impact the rail. The other technique applied was to angle the rail and have the shoe impact it in that manner.

In allowing the shoe/rail angle to be the maximum 0.14 deg, gouging did not occur. The resulting material response was very similar to Fig. 9.3. This indicates that the maximum angle allowed by the shoe geometry would cause wearing-type damage to the rail and some local elevated temperatures beyond the austenizing limit. Nearly identical results were obtained for an impact angle of 0.03 deg. Therefore, under normal sled operations (applied to the rail height tolerance and the allowable shoe gap scenario) only slight wearing will occur.

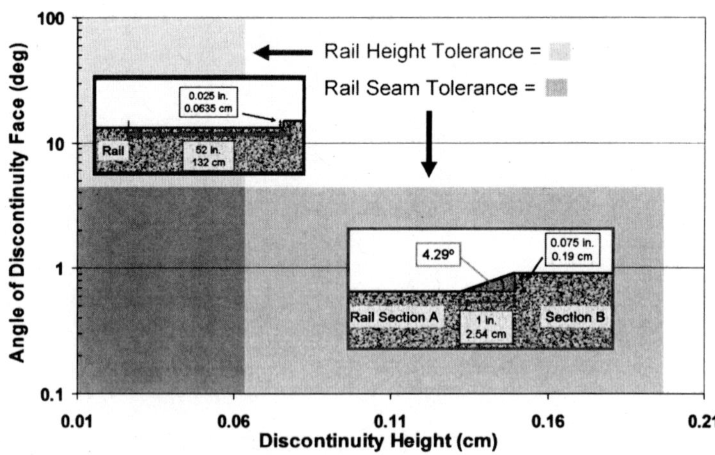

Fig. 9.19 Rail-alignment-tolerance summary.

B. Incidence Angle of 1.65 deg

If the angle of incidence were allowed to increase beyond what the shoe design would allow but within the tolerance allowed for a rail section seam, to an angle of 1.65 deg, the impact would also not lead to gouging. However, a very important effect was noted. Figure 9.20 illustrates the wearing- (or scraping) type damage to the rail and shoe during this type of impact. A small amount of plasticity is being generated, and the pressure is correspondingly indicated in Fig. 9.21.

Of particular interest is the temperature profile generated by this wearing-type impact. Figure 9.22 depicts the temperature generated by the small amount of plasticity during this kind of nongouging contact.

When we compare this generated heat profile from a nongouging, wear-type impact, to those in-service rails (with no gouges) analyzed in Chapter 4, we arrive at a possible explanation for the observed microstructural changes. Because the scrape reported on the in-service rail specimen designated ISRB was not measured and then painted over with coating, we cannot compare the material removal to that indicated in this simulation. However, that depth of austenization within the rail section matches very closely with that observed experimentally. Figure 9.23 compares the heat profile in this impact to the metallurgical analysis of specimen

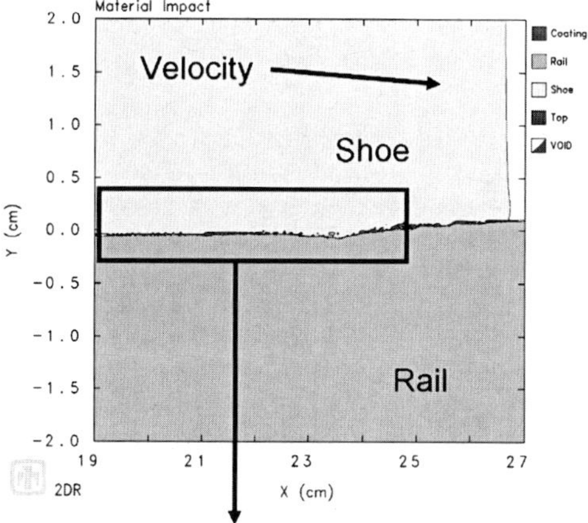

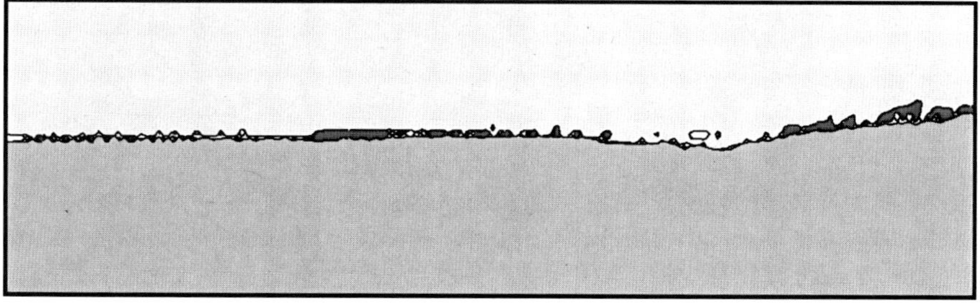

Fig. 9.20 Sled, 1.65-deg impact at 20 μs.

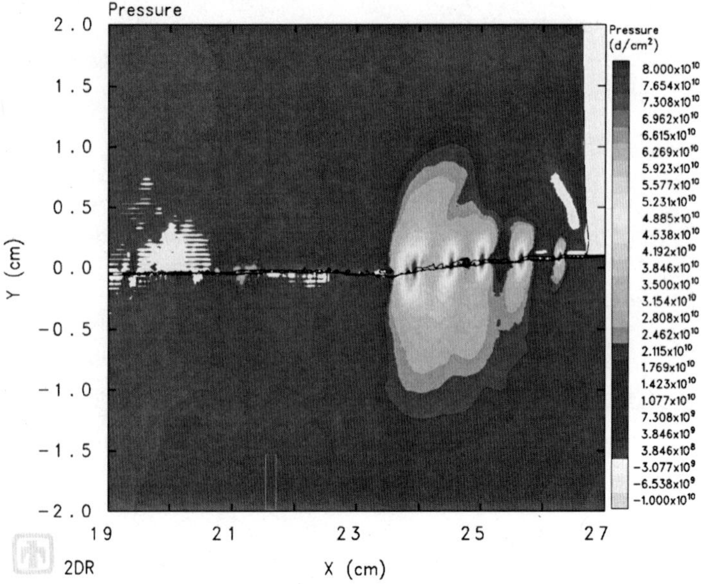

Fig. 9.21 Pressure, 1.65-deg impact at 20 μs.

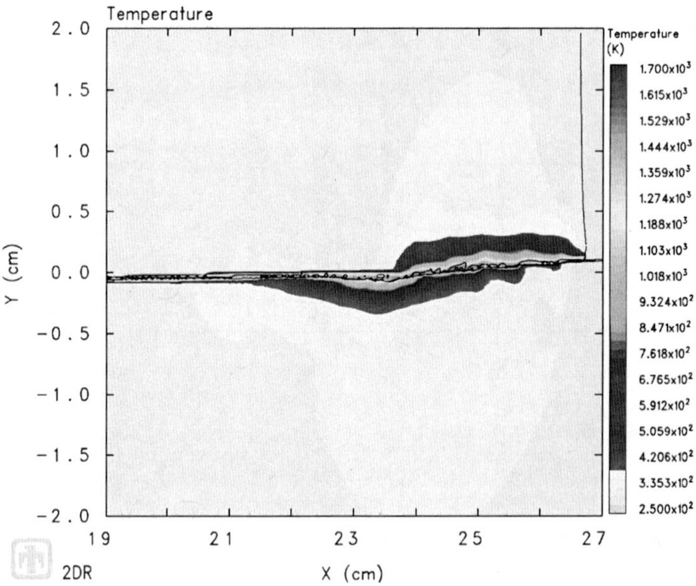

Fig. 9.22 Temperature, 1.65-deg impact at 20 μs.

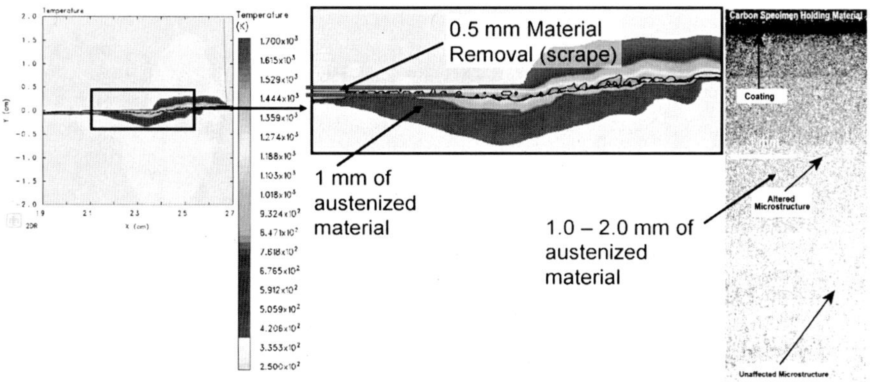

Fig. 9.23 Temperature profile comparison, 1.65-deg impact at 20 μs.

ISRB (which sustained a slight gouge and remained in service at the HHSTT), which appears in Fig. 4.24.

This indicates that a small, measurable amount of plasticity that was encountered in the field has caused a thermal pulse to enter the rail and change the microstructure of the rail steel. CTH is able to model this temperature effect, which is confirmed by metallurgical analysis of the in-service rails sent from the HHSTT.

C. Incidence Angle of 3 deg

To ascertain the conditions at which gouging would occur in the angled impact case to the extent it might result in the type of gouging analyzed metallurgically in Chapter 4, the incidence angle was increased to 3 deg. This might occur if the rear shoes catastrophically failed and allowed the rear sled travel to rise to 13 cm above the rail and the sled came down at this angle on the front shoes. Alternately, this angle is also within the tolerance limit for the sloping discontinuity at the rail section seams.

Figure 9.24 illustrates the resulting deformation and pressure resulting from this kind of impact. The generated temperature profile is depicted in Fig. 9.25. A comparison is made between this temperature profile and the experimentally examined gouge from the HHSTT in Fig. 9.26. Note that the magnitudes of material removal and the depth of austenization are very close to those observed metallurgically.

D. Summary of Angled Impact Case

Gouging was not generated at the maximum incidence angle, which can be expected during normal sled operations at the HHSTT when applied to the rail height tolerance. However, at angles within the allowable tolerances at the rail section seams, it was demonstrated that an impact at this angle could either create rail damage (wear and/or scraping), which could cause the microstructural changes seen in the nondamaged in-service HHSTT rails, or generate the kind of gouging impact that created the experimentally examined rail gouge that appears in Chapter 4.

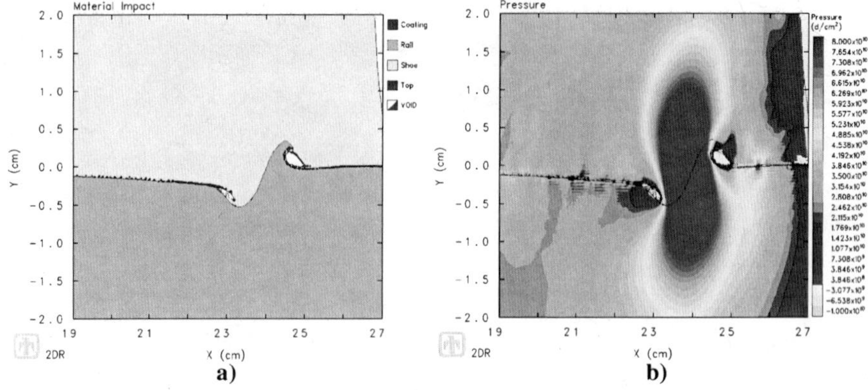

Fig. 9.24 a) Deformation and b) pressure, 3-deg impact at 20 μs.

These simulations increase our confidence in CTH modeling of hypervelocity impact and aid in determining which parameters in the HHSTT can lead to gouging.

V. Gouging Case 3: Rail Discontinuity

The final gouging case examined was the impact against, or into, a rail discontinuity. These continuities can take the form of an abrupt change in rail height, a smooth variation in rail height, or debris on the track (an asperity) [1]. This case is distinguishable from the angled impact scenario in that the rail discontinuity is

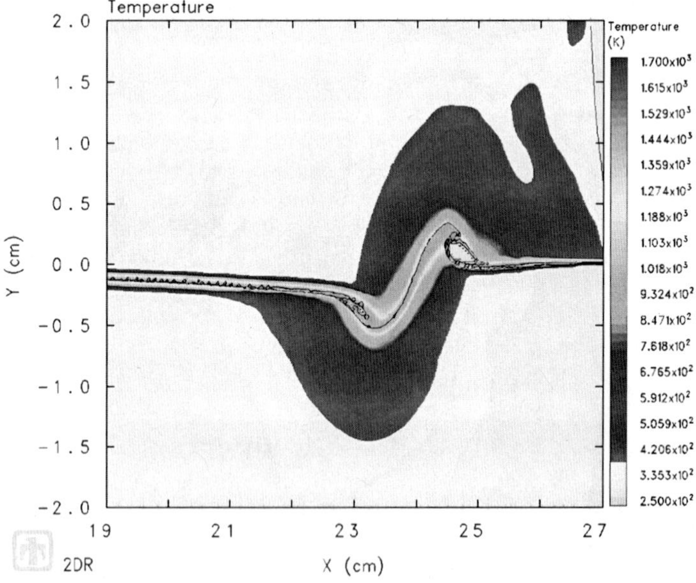

Fig. 9.25 Temperature, 3-deg impact at 20 μs.

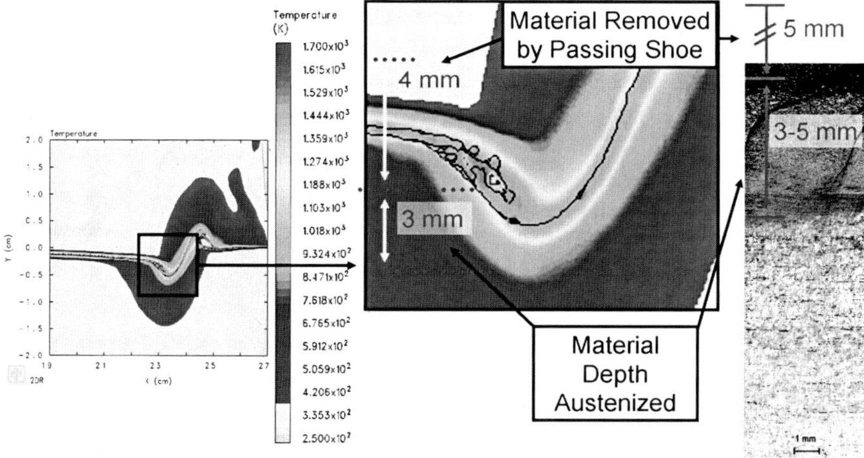

Fig. 9.26 Temperature comparison, 3-deg impact at 20 μs.

not a continuously angled rail, but a discrete section of rail that has a local rail height change.

In the modeling presented in this study, the type of rail discontinuity considered is a rail height change that might occur at the end of a rail section or might be inherent in the rail's condition. In all cases, the rail is considered to be coated with epoxy over its entire surface, including the discontinuity. In addition, the sled also has 1 m/s of vertical impact velocity. Of the cases examined, this is the most likely to occur in the field at the HHSTT. The allowable rail alignment tolerance used at the HHSTT is 0.0635 cm (Hooser, Michael, "HHSTT Track Tolerances," unpublished data from Holloman AFB, NM, 2006).

Initially, the discontinuities considered were "sharp" discontinuities, with a "face" angle of 90 deg. Figure 9.27 provides an overview of this test geometry. A discontinuity is placed in front of the sled and modeled to be coated.

As the study progressed, it became clear that altering the face angle would have a dramatic impact on the occurrence of gouging. Figure 9.28 illustrates the terminology adopted to examine the rail-discontinuity case more completely.

Based on the rail-alignment tolerances of 0.0635 cm over a span of 132 cm, the rail discontinuity would be a gradual slope. However, the method of checking for these rail height changes is a machine that runs down the track making measurements. This technique does not specifically flag local discontinuities as special cases. That is, the 0.0635-cm height change could occur in a sharp discontinuity, and it would still be within track tolerances. Based on the results of the angled impact, a very gradual face angle would not generate gouging. Therefore, the sharp discontinuities (with face angles of 90 deg) were considered as "worse-case" scenarios and were the subject of the first series of simulations.

Following an examination of these sharp discontinuities, a much more extensive series of simulations was conducted to explore the effect of face angle to

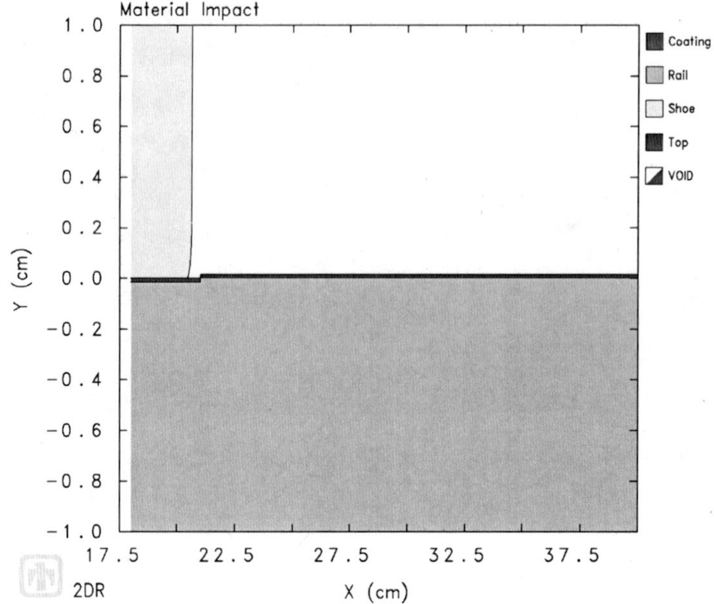

Fig. 9.27 Overview of sharp rail discontinuity.

the generation of gouging. Of course, the goal of this study was to discover the approaches required at the HHSTT to mitigate the occurrence of gouging.

A. 0.01524-cm Rail Discontinuity

The first case examined is a rail discontinuity of 0.01524 cm, which is the nominal thickness (6 mils) of the coatings placed on the rails [1]. This specific case appears in Fig. 9.27. This small discontinuity was compressed into the rail, and, although it caused some wearing-type damage to the shoe and rail, it did not initiate gouging. The deformation and temperature profile that CTH generates as the solution to this impact problem appears in Fig. 9.29. This degree of discontinuity does not generate gouging, but does results in some localized plasticity and temperature development.

This amount of temperature generation is similar in magnitude to that seen on the nondamaged in-service rails examined in Chapter 4. Specifically, specimen

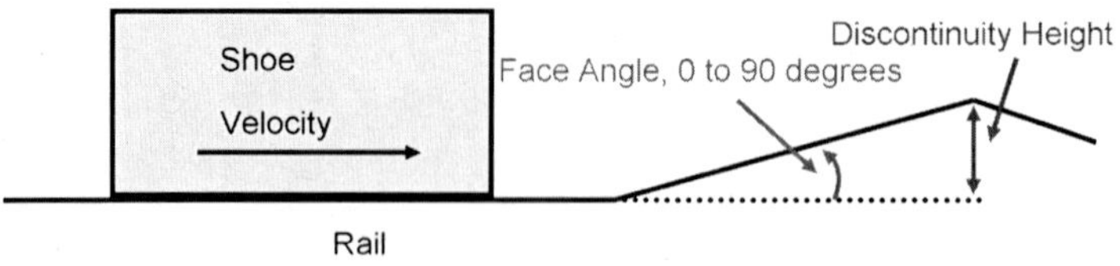

Fig. 9.28 Overview of general rail discontinuity.

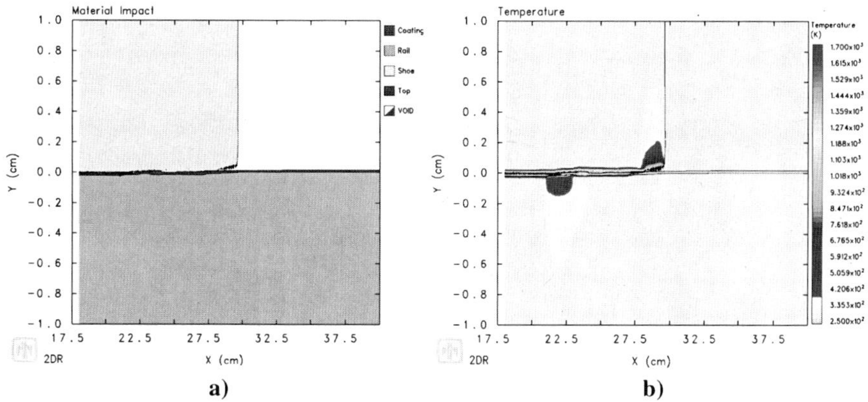

Fig. 9.29 Sled, 0.01524-cm rail discontinuity at 20 μs, a) deformation and b) temperature.

"IS" (see Fig. 4.23) is depicted in Fig. 9.30 along with the temperature profile created by this 0.01524-cm discontinuity impact. This provides another explanation for the microstructure changes seen in the rails that appear (visually) not to have any damage.

B. 0.02-cm Rail Discontinuity

In an effort to find a discontinuity that would initiate gouging, a 0.02-cm discontinuity was placed into the model, similar to that depicted in Fig. 9.27. This level of discontinuity led to the creation of a typical gouging impact. Figure 9.31 depicts the resulting deformation and pressure plots for this case. Figure 9.32 indicates the plastic strain-rate and temperature results. Note that the strain-rate reaches the 10^6/s level in some areas, and throughout the specimen the full range of strain rates is being experienced. Additionally, the temperature plot indicates a portion

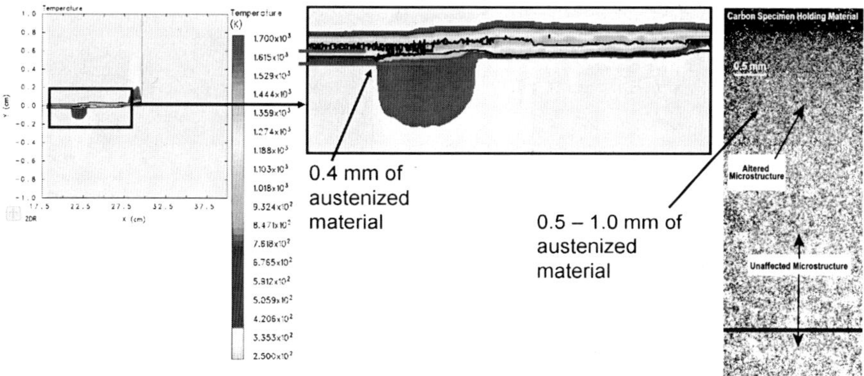

Fig. 9.30 Comparison of temperature profile, 0.01524-cm rail discontinuity at 20 μs.

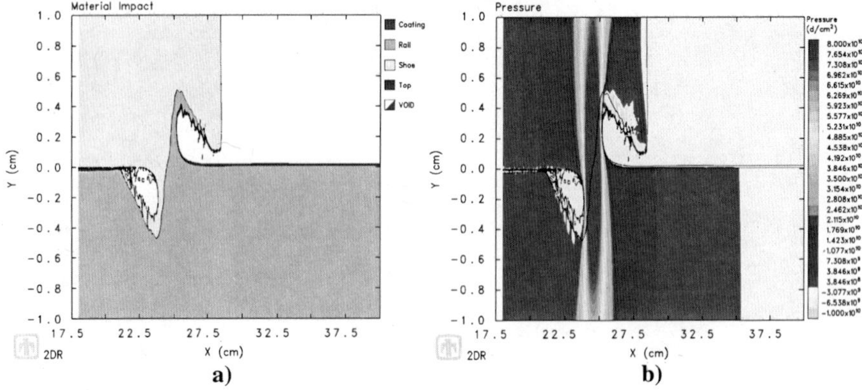

Fig. 9.31 Sled, 0.02-cm rail discontinuity at 25 μs, a) deformation and b) pressure.

of the material (shown in red) has reached the melt temperature (~1650 K). This matches more completely the experimental findings in the gouged specimen from the HHSTT, both in this work, and in Gerstle [17]. In the other gouging cases in this work, the melting temperature was not attained. This particular CTH simulation was allowed to run out to 25 μs to capture more of the event.

This 0.2-cm discontinuity case compares well against the experiment gouge analysis presented in Chapter 4. Figure 9.33 illustrates how the major elements of that gouge are matched with this simulation to include melting at the gouge surface.

C. 0.03048-cm Rail Discontinuity

The case of a discontinuity at twice the coating thickness was examined to further explore the phenomenon of gouging. Figure 9.34 illustrates the deformation and pressure plots, whereas Fig. 9.35 depicts the strain rates and temperature. This

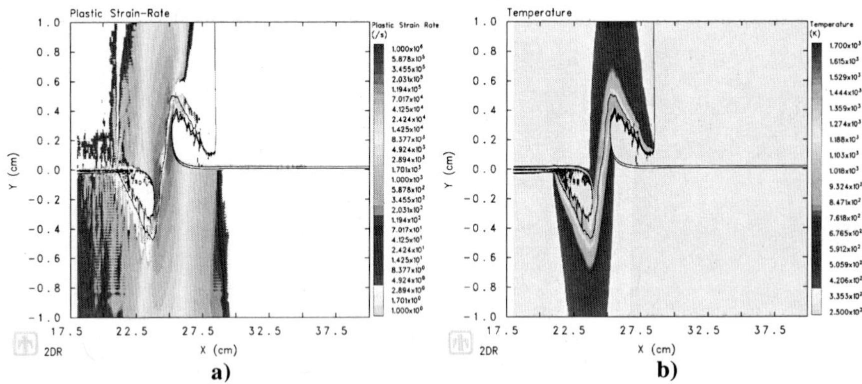

Fig. 9.32 Sled, 0.02-cm rail discontinuity at 25 μs, a) strain-rate and b) temperature.

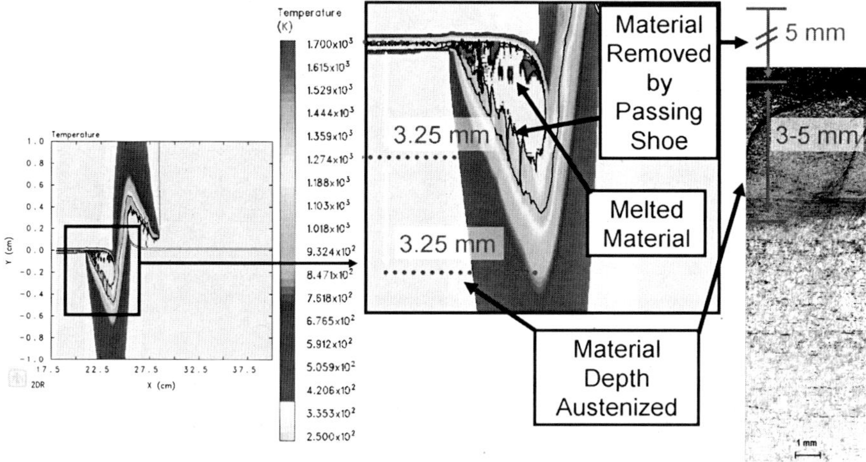

Fig. 9.33 Temperature profile comparison, 0.02-cm rail discontinuity at 25 μs.

particular simulation was stopped at 20 μs to compare to the other types of gouging impacts examined in this work.

Again, the strain rates are focused in the high regime, but significant portions of the plasticity are occurring in the midrange strain rates, making the full-range constitutive model important to an accurate solution. The temperature profile in this case compares well with the metallurgical gouging evidence (see Fig. 9.36). Note that the deformations are not more pronounced because of stopping the simulation 5 μs before the 0.2-cm discontinuity case.

Of course, allowing the simulation to run further in time might create a profile that exactly matched the experimental gouge. However, the creation of these very close matches between the CTH simulation and the experimental record allows the conclusion that the model is generating accurate results.

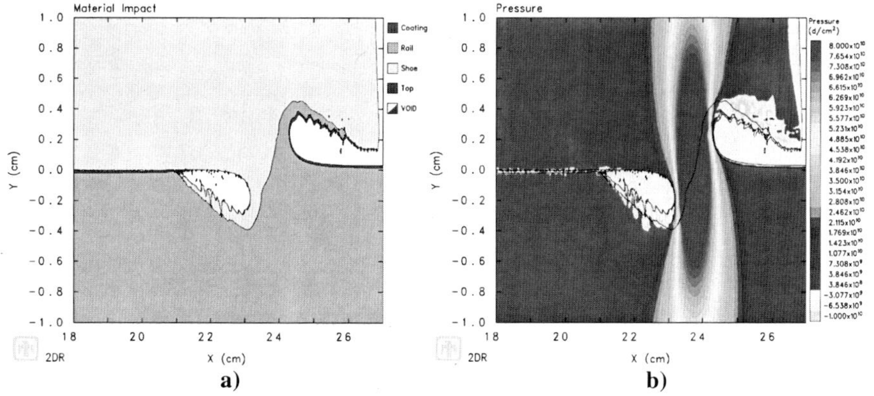

Fig. 9.34 Sled, 0.03048-cm rail discontinuity at 20 μs, a) deformation and b) pressure.

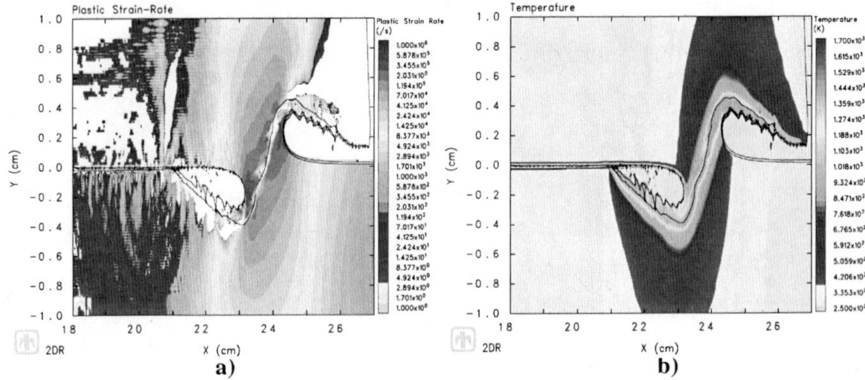

Fig. 9.35 Sled, 0.03048-cm rail discontinuity at 20 μs, a) strain rate and b) temperature.

D. Discontinuity with Varying Face Angle

Noting that the sharp discontinuity was one extreme version of this case, and that the angled impact could be viewed as the other extreme (that is, a discontinuity that occurred over an extended distance), an in-depth evaluation of numerous impact conditions was conducted. The goal was to find the combination of the discontinuity height and face angle (reference Fig. 9.28) that created a gouging impact, or more importantly, to define the set of parameters that would lead to a wearing-type impact and allow the sled to pass without initiating gouging (and perhaps catastrophic failure).

To illustrate the type of impact that was created, a representative example is presented here. Figure 9.37 depicts an impact with a 0.0635-cm discontinuity, which has a 2-deg face angle. This rail condition is within both rail-alignment

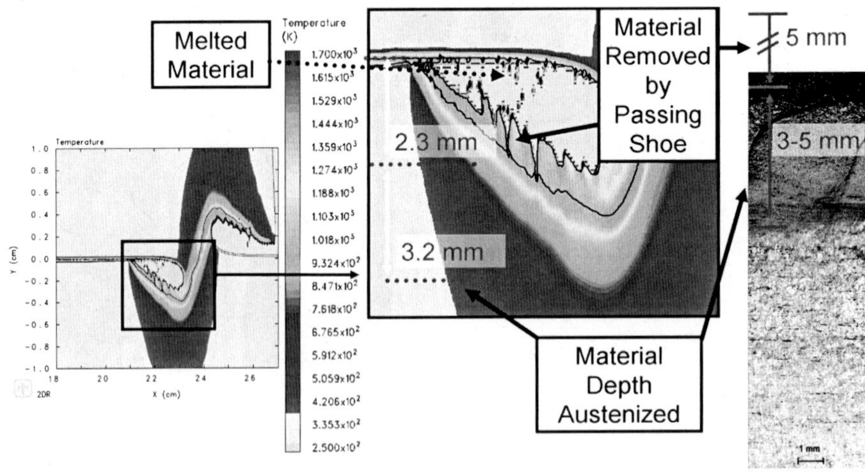

Fig. 9.36 Temperature profile comparison, 0.03048-cm rail discontinuity at 20 μs.

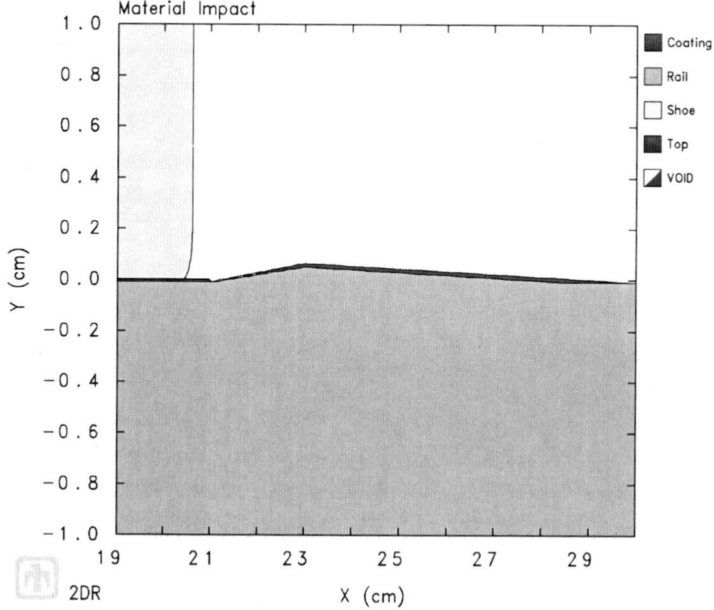

Fig. 9.37 Example of discontinuity with an angled face.

tolerances of total rail height change over an extended distance and the local tolerance for rail seams.

As the impact progresses, the characteristic hump is created in front of the shoe, and material mixing begins. This leads to the gouging-type material interaction. Figure 9.38 shows this material flow beginning.

The type of gouging that results as the impact event unfolds closely resembles those presented earlier in this chapter. Figure 9.39 shows the deformation and pressure at 40 μs. The characteristic high-pressure core is clearly evident. The strain rates and temperatures generated by the impact are depicted in Fig. 9.40. Once again, the strain rates are throughout the entire range, and there is a zone of austenized steel below the gouge.

This gouging case also compares well against the experimental gouge characteristics presented in Chapter 4. Figure 9.41 illustrates this comparison.

Based on these findings (i.e., that gouging was affected by the face angle as well as the discontinuity height), a full complement of CTH simulations was conducted to define the parameter values that would only lead to a wearing-type damage. Table 9.2 summarizes the results of this extensive study.

This information can be graphically presented in a manner that shows a distinct relationship between the parameters and the occurrence of gouging. Figure 9.42 shows these cases on a plot of discontinuity height vs face angle. The gouge and wear results are shown, with a threshold shown between the gouging and wearing damage cases. Because no gouging occurred below a face angle of 1.85 deg, that value is plotted using a dashed line. Additionally, a power-law fit is made to the wear data points and the threshold values until the angle decreases to the threshold gouging angle of 1.85 deg. To highlight the information at small angles, Fig. 9.43 shows the same information, with the face angle depicted on a logarithmic scale.

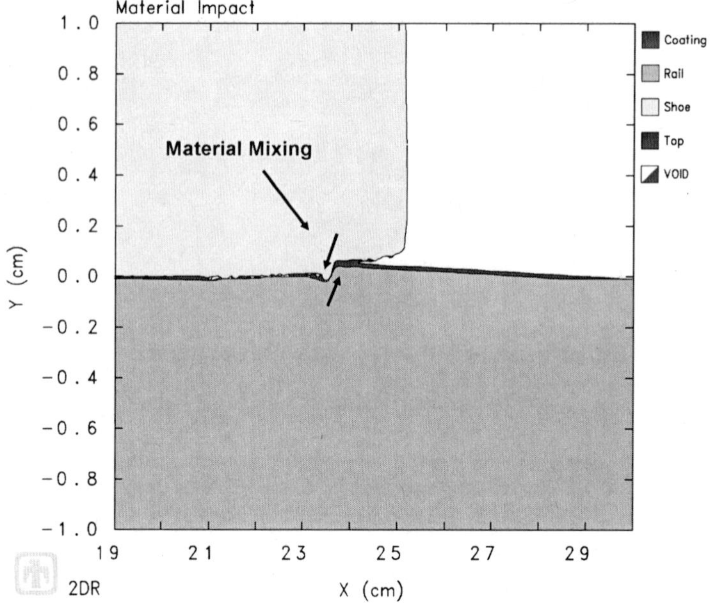

Fig. 9.38 Example of discontinuity with an angled face at 15 μs.

Placing the current HHSTT tolerances for overall rail-height variation and rail-section seam discontinuity on a plot of the simulation results shows that the HHSTT is operating in the area of predicted gouging. This is illustrated in Fig. 9.44. The area defined by a discontinuity below 0.0635 cm in magnitude (at any face angle) is the current rail-height tolerance. The area defined by an angle below 4.3-deg face angle and a discontinuity less than 0.19 cm represents the current rail-seam tolerance. Therefore, the HHSTT operates in these areas on Figure 9.44 that the CTH simulations show lead to hypervelocity gouging.

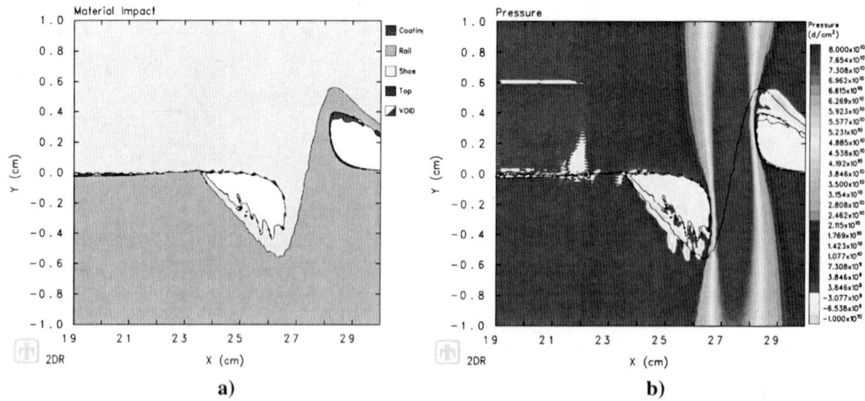

Fig. 9.39 Example of discontinuity with an angled face at 40 μs, a) deformation and b) pressure.

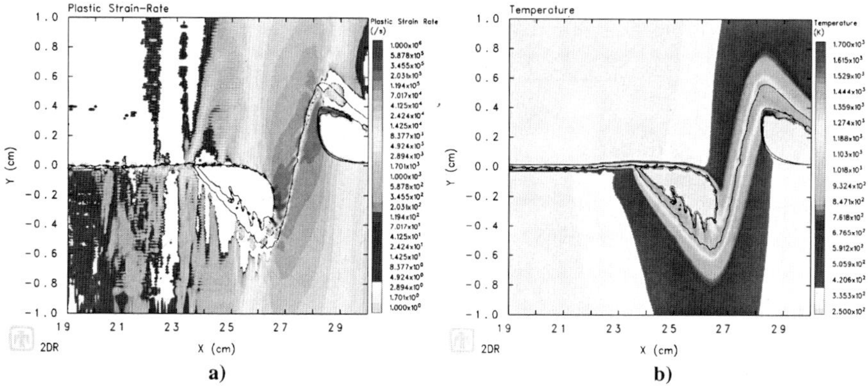

Fig. 9.40 Example of discontinuity with an angled face at 40 μs, a) strain rate and b) temperature.

This clearly indicates that the gouging experiences at the HHSTT can occur when the rail is prepared per their current rail-alignment tolerances. Therefore, in order to prevent gouging in future sled run, the alignment criteria must be adjusted to remain within the zone of predicted wear. Expressed in an equational sense, and using the wear power fit results, the new criteria would be

$$
\begin{aligned}
&\text{Face Angle (deg)} = \max[(9.174 \cdot 10^{-5} \cdot \text{Discontinuity (cm)}^{-3.309}), 1.85^\circ] \\
&\text{Face Angle (deg)} = \max[(4.196 \cdot 10^{-6} \cdot \text{Discontinuity (in)}^{-3.309}), 1.85^\circ]
\end{aligned}
\tag{9.1}
$$

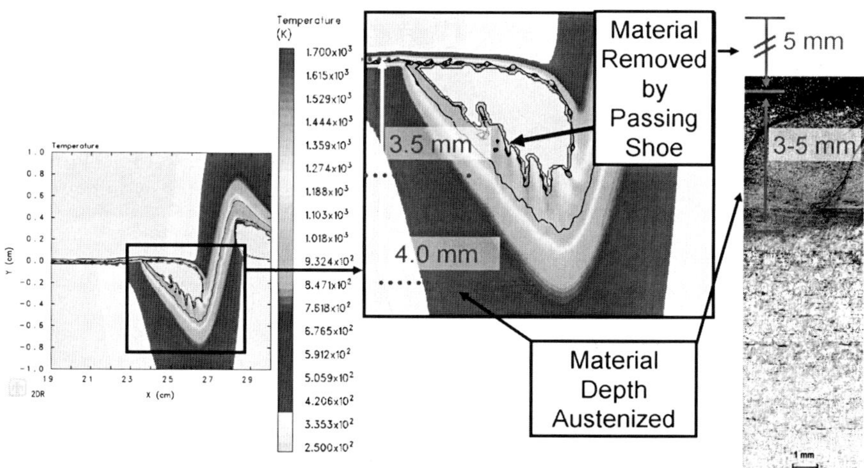

Fig. 9.41 Example of discontinuity with an angled face at 40 μs, temperature profile comparison.

Table 9.2 Summary of discontinuity with an angled face simulation results

Face angle, deg	Discontinuity, cm	Damage
0.014	0.07	Wear
1.6	0.055	Wear
1.6	0.06	Wear
1.65	0.07	Wear
1.85	0.0635	Wear
1.85	0.07	Wear
2	0.05	Wear
2	0.06	Gouge
2	0.0635	Gouge
2	0.07	Gouge
3	0.0635	Gouge
3	0.07	Gouge
4	0.07	Gouge
5	0.035	Wear
5	0.045	Gouge
10	0.03	Wear
10	0.035	Gouge
20	0.025	Wear
20	0.03	Gouge
20	0.05	Gouge
20	0.635	Gouge
40	0.02	Wear
40	0.025	Gouge
60	0.018	Wear
60	0.022	Gouge
60	0.035	Gouge
90	0.015	Wear
90	0.02	Gouge
90	0.03	Gouge
90	0.0635	Gouge

That is, the maximum allowable face angle, as a function of discontinuity height, is expressed in the preceding. This rule can be used as the new rail-alignment tolerance limit, which, according to the CTH simulation predictions, would prevent the occurrence of hypervelocity gouging.

E. Summary of Rail Discontinuity Case

In this particular gouging case, the hypervelocity gouging was initiated by discontinuities in the rail height. This is a very plausible condition at the HHSTT. Note that when the rail-alignment tolerances are used at the HHSTT, gouge initiation can occur within those tolerances. Additionally, the character of the gouging was a better fit to the experimental evidence presented in Chapter 4 because the material in the gouge reached the melting temperature.

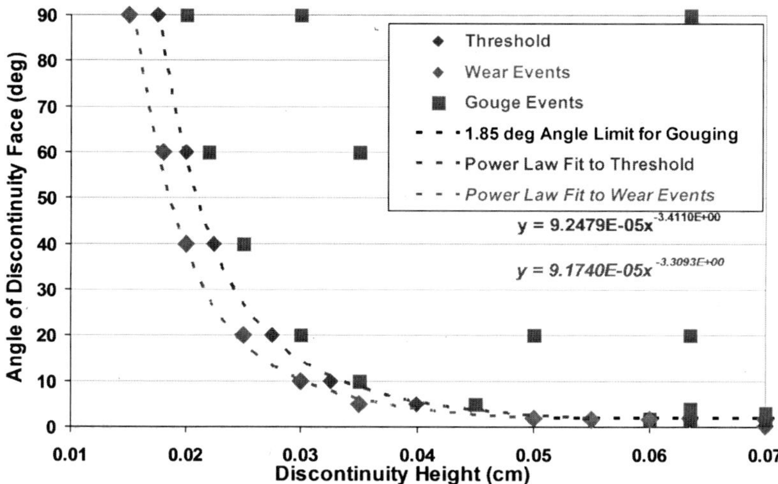

Fig. 9.42 Discontinuity with an angled face simulation results.

VI. Conclusions

In this chapter, three different techniques were employed in the CTH model of the HHSTT hypervelocity gouging scenario to initiate the gouging process. The results of this study were the identification of key aspects of gouging, which was evident in the metallurgical examination of rail gouges and other rail sections (see in Chapter 4).

The first was the vertical impact velocity approach. Using this technique, only when the vertical velocity was increased to levels much higher than those present in the HHSTT scenario did gouging occur. Throughout a majority of the cases,

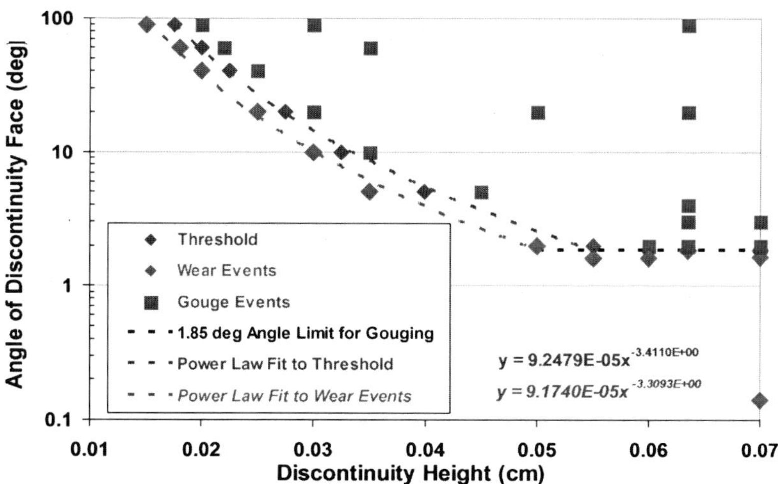

Fig. 9.43 Discontinuity with an angled face simulation results (logarithmic scale).

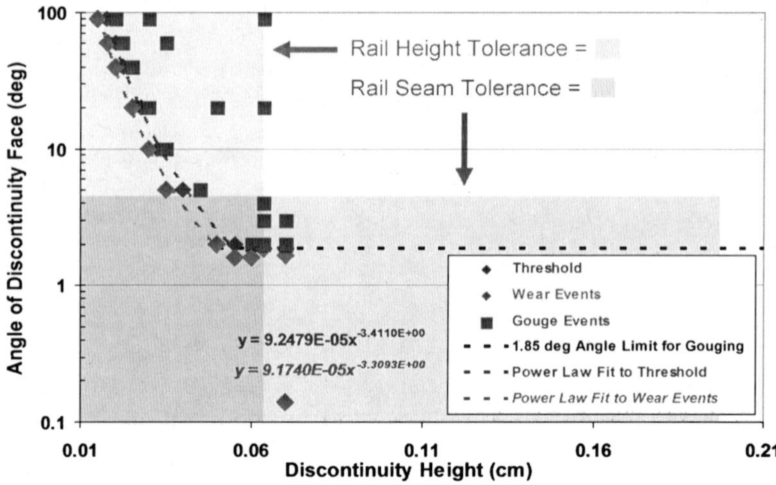

Fig. 9.44 Discontinuity with an angled face simulation results vs HHSTT tolerances.

the one-dimensional penetration theory predicted the resulting penetration depth fairly well. In the impacts that did not gouge, localized plasticity generated temperature above the austenizing temperature, but not down to the depths observed in the rail samples examined in Chapter 4. In the impacts that gouged, assuming sufficient vertical velocity was applied, the characteristics of the gouging matched the metallurgically examined gouge from the HHSTT. A heated shoe was also modeled, which demonstrated very little difference from the unheated shoe case.

The second technique was to induce an angled impact. When the maximum allowable angle (i.e., possible without shoe structural failure) was applied to a flat rail, no gouging occurred. However, if the rail were angled within the tolerances allowable for rail alignments at the section seams, a wearing-type impact phenomenon generated an austenizing temperature profile down to 1 mm into the rail, which matches fairly closely to the in-service rail sections examined in Chapter 4. Hypervelocity gouging was created by increasing the incidence angle beyond the nominal maximum for a flat rail, but also within the seam tolerances at the HHSTT. This gouge also matched well to the HHSTT gouge.

The final approach used was to establish a rail discontinuity and allow the shoe to run into it. Rail discontinuities well below the rail-alignment tolerances created gouging. However, at the discontinuity value equal to the coating thickness, no gouging occurred, but a wearing-type impact generated austenizing temperatures that could explain the microstructural variations observed in the in-service rails examined in Chapter 4. When the discontinuities were increased, gouging on the order of that evaluated from the HHSTT gouge was observed, including material melting and austenizing temperatures into the depth of the rail that matched measured values. Additionally, an extensive study of various discontinuities and face angles resulted in the description of a gouging limit that could be used to prevent gouging at the HHSTT.

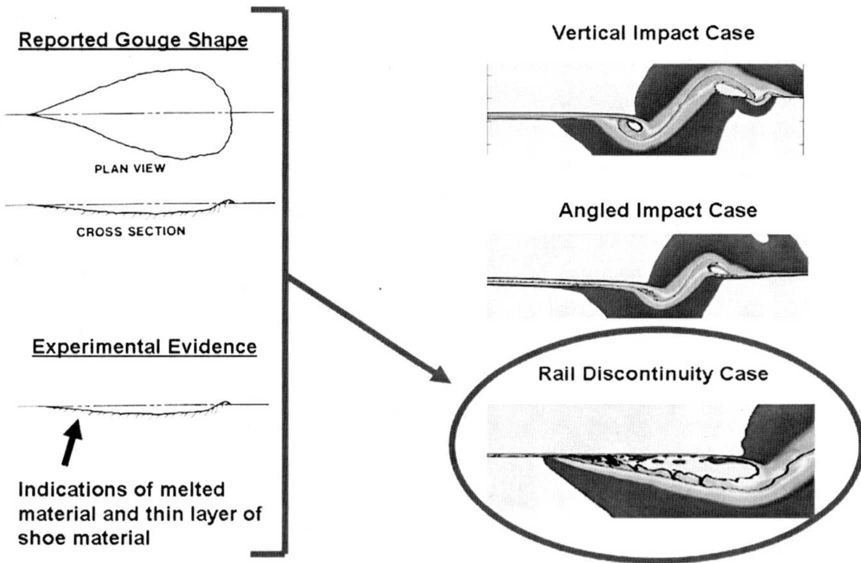

Fig. 9.45 Comparison of gouge characteristics.

In making a judgment concerning which scenario is the most plausible, aside from the fact that the gouging only occurs in the first case when the simulation conditions exceed the normal limits of a sled run, one can look at the gouges themselves. In comparing the shape of the generated gouge, the amount of shoe material deposited in the rail, and the creation of melted material in the gouge, the rail discontinuity case creates better results. Figure 9.45 illustrates this comparison. The gouges represent the best matches from each case to the HHSTT gouge (from Chapter 4) and have only been altered to make the gouges in the correct aspect ratio and the same length in the diagram.

The first two techniques generate gouges that leave too much shoe material in the rail material and that do not match the gouge shape reported in the literature. The rail discontinuity case, however, matches all aspects of the experimentally examined gouge more closely, including the shape.

Based on the study performed in this chapter, it is clear that the gouges being created during the HHSTT sled runs are most likely rail discontinuity induced. These gouges occur with the validated CTH model, using the newly developed material constitutive models and rail alignments below the HHSTT tolerances. To avoid gouging on the rail, the discontinuities must be reduced to below the limits described in Eq. (9.1).

References

[1] Szmerekovsky, A. G., "The Physical Understanding of the Use of Coatings to Mitigate Hypervelocity Gouging Considering Real Test Sled Dimensions AFIT/DS/ENY 04-06," Ph.D. Dissertation, Department of Aeronautics and Astronautics, Air Force Inst. of Technology, Wright-Patterson AFB, Dayton, OH, Sept. 2004.

[2] Cinnamon, J. D., and Palazotto, A. N., "Further Refinement and Validation of Material Models for Hypervelocity Gouging Impacts," *AIAA Journal*, Vol. 46, No. 2, 2008, pp. 317–327.

[3] Cinnamon, J. D., and Palazotto, A. N., "Analysis and Simulation of Hypervelocity Gouging Impacts for a High Speed Sled Test," *International Journal of Impact Engineering*, Vol. 36, 2009 (to be published).

[4] Laird, D. J., "The Investigation of Hypervelocity Gouging AFIT/DS/ENY 02-01," Ph.D. Dissertation, Department of Aeronautics and Astronautics, Air Force Inst. of Technology, Wright-Patterson AFB, Dayton, OH, March 2002.

[5] Szmerekovsky, A. G., and Palazotto, A. N., "Structural Dynamics Considerations for a Hydrocode Analysis of Hypervelocity Test Sled Impacts," *AIAA Journal*, Vol. 44, No. 6, 2006, pp. 1350–1359.

[6] Szmerekovsky, A. G., Palazotto, A. N., and Baker, W. P., "Scaling Numerical Models for Hypervelocity Test Sled Slipper-Rail Impacts," *International Journal of Impact Engineering*, Vol. 32, No. 6, 2006, pp. 928–946.

[7] Szmerekovsky, A. G., Palazotto, A. N., and Ernst, M. R., "Numerical Analysis for a Study of the Mitigation of Hypervelocity Gouging," AIAA Paper 2004-1922, April 2004.

[8] Szmerekovsky, A. G., Palazotto, A. N., and Ernst, M. R., "Numerical Analysis for a Study of the Mitigation of Hypervelocity Gouging," *Dayton-Cincinnati Aerospace Science Symposium*, modified and published as AIAA Paper 2004-1922, March 2004.

[9] Szmerekovsky, A. G., Palazotto, A. N., and Cinnamon, J. D., "An Improved Study of Temperature Changes During Hypervelocity Sliding High Energy Impact," AIAA Paper 2006-2090, May 2006.

[10] Laird, D. J., and Palazotto, A. N., "Effects of Temperature on the Process of Hypervelocity Gouging," *AIAA Journal*, Vol. 41, No. 11, 2003, pp. 2251–2260.

[11] Laird, D. J., and Palazotto, A. N., "Temperature Effects on the Gouging and Mixing of Solid Metals During Hypervelocity Sliding Impact," AIAA Paper 2002-1691, April 2002.

[12] Blomer, M. A., "An Investigation for an Optimum Hypervelocity Rail Coating AFIT/GAE/ENY/05-J01," Master's Thesis, Department of Aeronautics and Astronautics, Air Force Inst. of Technology, Wright-Patterson AFB, Dayton, OH, May 2005.

[13] Cinnamon, J. D., Palazotto, A. N., and Brar, N. S., "Further Refinement of Material Models for Hypervelocity Gouging Impacts," AIAA Paper 2006-2086, May 2006.

[14] Cinnamon, J. D., Palazotto, A. N., Szmerekovsky, A. G., and Pendleton, R. J., "Further Investigation of a Scaled Hypervelocity Gouging Model and Validation of Material Constitutive Models," AIAA Paper 2006-2087, May 2006.

[15] Kennan, Z., "Determination of the Constitutive Equations for 1080 Steel and Vascomax 300 AFIT/GAE/ENY/05-J05," Master's Thesis, Department of Aeronautics and Astronautics, Air Force Inst. of Technology, Wright-Patterson AFB, Dayton, OH, May 2005.

[16] Hooser, M., "Simulation of a 10,000 Foot Per Second Ground Vehicle," AIAA Paper 2000-2290, June 2000.

[17] Gerstle, F. P., Follansbee, P. S., Pearsall, G. W., and Shepard, M. L., "Thermoplastic Shear and Fracture of Steel During High-Velocity Sliding," *Wear*, No. 24, 1973, pp. 97–106.

Chapter 10

Conclusions

I. Introduction

THIS study examined the phenomenon of hypervelocity gouging and resolved several of the limitations of previous efforts in the field. Through a rigorous experimental process, material models were developed and validated against observed results. A computational model that captures the nonlinear nature of the constitutive models and the nonequilibrium thermodynamic nature of the equation of state was created. A full-range strain-rate material flow model was shown to generate results that match experimentation. This model was applied to the HHSTT gouging problem, and observations were made on the origin of the damage and nature of its development [1, 2].

II. Review of Previous Research in the Field

A summary of the past research efforts in the field of hypervelocity gouging revealed several shortcomings and illuminated the direction that further investigations should take. To span the wide number of approaches to this phenomenon, this study acknowledged both the analytical and experimental nature of the endeavor and successfully joined them in a single work.

One of the primary avenues available in the effort to analyze and model hypervelocity gouging is the experimental evidence from the HHSTT. Although instrumenting a gouge while it occurs or purposely creating a gouging impact is not possible, many experimentally verifiable aspects of the phenomenon can be utilized. This study, then, sought to examine the gouging problem using state-of-the-art laboratory equipment and techniques.

The inability to generate instrumented or intentional gouging on the sled track had led past investigators into the creation of laboratory experiments to create gouging. Based on the valuable nature of having a hypervelocity gouge created under measurable and controlled condition, this study created a hypervelocity gouging experiment to use as a validation of modeling efforts.

Past modeling efforts in the field have been extensive. Although the most recent efforts have established an optimum computational code choice (CTH) and resulted in the simulation of gouging within the code solution, the codes were relying on inexact material constitutive models for the materials in the HHSTT scenario. Additionally, the material model for VascoMax 300 was a strain-rate independent implementation. Finally, the contact algorithms utilized were not the optimum choice for numerical stability. This study corrected those deficiencies by determining an experimentally based material flow model for each material and validating the models with a CTH code simulation of impact events.

In this way, the review of past research efforts established the areas to focus this study in order to generate the most accurate results possible. The remainder of this chapter will summarize each of these specific areas and highlight the conclusions available from them.

III. Theoretical Foundations

The theoretical foundation of this research study revolves around the manner in which the hypervelocity gouging problem can be modeled within a computational code. Central to this approach is the understanding of how these types of problems are solved.

The solution of this type of scenario within the CTH code involves dividing the problem into two regimes. In the high-pressure and -temperature portion of the deformation event, the material equation of state dominates the solution. CTH is able to model this in a uniquely capable manner by relying on experimentally determined material states within the code. These values capture all of the effects present in high-energy deformation, including the various nonequilibrium thermodynamic effects. The EOS models used were specific to VascoMax 300 and 1080 steel.

Within the regime that is dominated by the material strength and constitutive models, the specific flow formulation is the essential element. Two primary strain-rate dependent formulations that can be created from experimentation are available within CTH.

Finally, the material failure model was summarized. The choice selected was the most commonly used, and most accurate, for metals undergoing nonexplosive deformation.

IV. Gouge Characterization

Using experimental techniques not applied to the HHSTT gouging phenomenon previously, a gouge was rigorously examined to understand the nature of the deformation process. Additionally, various other sections of rail were examined to better understand the conditions at the HHSTT.

A hypervelocity gouge was sectioned and examined through the depth to ascertain the key aspects of hypervelocity gouging. It was immediately evident that the material had experienced a thermal pulse that had changed the material microstructure during the postgouge cooling. Various state-of-the-art techniques were applied to the specimen, and a thermal profile was determined.

The experimentally observed thermal profile was modeled using a one-dimensional heat conduction model. The model, based on the heat-conduction capabilities of 1080 steel, was able to replicate the observed microstructure. Therefore, a verified thermal history was established for the gouged specimen. This served to provide a set of conditions to validate computational simulations against.

In addition to examining a hypervelocity gouge, other rail sections from the HHSTT were examined in a similar manner. For rail sections that had never been placed in service, the manufacturing technique of head hardening was discovered to be taking place. On the rail section that had been in service at the HHSTT, microstructural changes that were similar (although to a lesser extent) to the gouged specimen were noted. One section had a "wear"-type impact that "scraped" the rail and was later painted over with coating. The other two sections had no visible damage. All of these in-service sections showed the same kind of thermal pulse evidence.

This examination of the rail at the HHSTT, including a hypervelocity gouge and a wearing impact, served to create validation points to match computational simulations against.

V. Development of Material Constitutive Models

To create accurate material constitutive models for use in the gouge modeling process, a series of experimental tests were conducted. These were designed to investigate and determine the midrange strain-rate and high strain-rate material responses.

The split Hopkinson bar test (SHB) was employed to test the midrange (10^3/s) strain-rate material behavior. Both VascoMax 300 and 1080 steel were tested, and their material dynamic yield strength was computed. With these experiments it was possible to generate the constitutive models.

A high strain-rate experiment, the flyer impact plate, was conducted on the two materials to define the material response in the 10^4/s to 10^5/s strain-rate range. To generate the constitutive models, the flyer plate experiments were modeled within CTH, using its experimental equation-of-state models. The material flow models were determined in an iterative process that sought to match the stress curves measured in the flyer plate experiments. Any nonequilibrium thermodynamic effects present in the flyer experiments were being modeled by the particular equation-of-state model in CTH, which is based on high-energy experimentation and measurement.

To create material constitutive models that matched the high strain-rate experiments and also maintained a fit to the midrange data, the Zerilli–Armstrong model formulation was selected as the appropriate choice for use in CTH modeling.

VI. Validation of the Constitutive Models for Midrange Strain Rates

Although the hypervelocity gouging problem has areas that are at high strain rates and in the realm of equation-of-state computations, much of the solution occurs within the midrange strain-rate regime. Therefore, a study was conducted to validate the new material flow models at impact in the midrange regime.

To validate the SHB tests, a finite element code model was developed using ABAQUS®. This code was able to create the posttest geometry of the SHB specimens using the new Johnson–Cook constitutive model. (The code could not implement the Zerilli–Armstrong formulation.) At these strain rates, the Zerilli–Armstrong and the Johnson–Cook formulations are fairly close in the estimate of flow stress.

The SHB specimens were sectioned and metallurgically examined to ascertain if their microstructure matched the flow and shear band characteristics observed in the finite element model. Those elements were experimentally verified in the deformed microstructure.

To establish a midrange strain-rate impact experiment to validate CTH's capability to model accurately with these material models, a Taylor impact test was constructed. A series of tests with VascoMax 300 and 1080 steel projectiles were conducted and simulated within CTH. With the new material flow models, the accuracy of the results was one-half to one order of magnitude better than with the material models used by previous researchers. The material deformation was matched within 5%.

In a related effort, a series of Taylor tests were conducted with the nose of the projectile coated with the materials used at the HHSTT to mitigate gouging and another candidate coating. The impact results demonstrated that there was a relative difference in the coefficient of friction between the coatings. Epoxy was shown to be the most effective coating, especially when compared against an uncoated rail, reducing friction by 31%.

Simulations of the coated Taylor tests were conducted, and similarly good results were obtained for the coated specimens. In addition, the behavior of the coatings was replicated by CTH, validating the flow models and equation-of-state models for the coatings within the code.

The CTH simulation techniques, boundary conditions, and contact algorithms were discussed, based on a related study of those topics. The conclusion of that exploration was that the contact algorithm needed to be the default (no-slide) condition in CTH and that an approach used by previous investigators might lead to unstable simulation results.

VII. Development of a Scaled Hypervelocity Impact Experiment

A laboratory hypervelocity gouging test to fully validate the material flow models for modeling in CTH was desired. This hypervelocity test would serve to provide measurable data at known conditions that could be used to check the CTH simulations.

To this end, a study was performed to determine the mathematically scaled version of the HHSTT scenario within the laboratory environment. The Buckingham pi approach was used to scale the full problem. It was found that the gun range facility available was not able to conduct experiments of the velocity and geometry desired. Therefore, an optimum test configuration was arrived at by attempting to match the Buckingham pi parameters within the constraints of the test facility.

Based on the small number of tests available, an approach was needed to ensure gouging was created by the laboratory test. To estimate this configuration, a novel one-dimensional penetration theory was developed and applied to the gouging scenario. This theory is empirically grounded and provides a penetration depth estimate based on impact geometry. The theory was applied to the HHSTT sled gouging problem and demonstrated that it matched some of the known conditions that lead to gouging. The one-dimensional theory was then applied to the scaled hypervelocity gouging test to determine the required impact angle for the projectiles to create gouging.

The computed test configuration was utilized, and a series of hypervelocity impact tests were conducted. The one-dimensional penetration theory demonstrated remarkable agreement with the various types of impacts that were generated. Additionally, gouging was created by the laboratory test and accurately predicted by the one-dimensional theory.

Finally, the gouges created by the laboratory experiment were metallurgically examined to ascertain if the same kind of thermal history witnessed in the HHSTT gouge was observed. Indeed, the microstructure of the laboratory gouges contained the same kind of changes that had been seen in the HHSTT gouge.

The hypervelocity gouging tests, then, created gouges that could be used as validation points for the CTH simulation to achieve. In addition, the thermal profiles generated by CTH would be used to match against the observed metallurgical evidence.

VIII. Validation of the Constitutive Models for High Strain Rates

To ensure that CTH simulation results are accurate for modeling efforts in hypervelocity impact problems, the hypervelocity gouge tests were used as a validation.

A general discussion of plane-strain solutions to three-dimensional gouging problems was presented. This summarized a related study that demonstrated good agreement between a two-dimensional axisymmetric simulation in CTH to a two-dimensional plane-strain simulation conducted to the point in time in which reflected waves would affect the area of interest.

With this two-dimensional plane-strain implementation shown to be accurate, the laboratory hypervelocity gouge tests were modeled within CTH using the material flow models developed in this study. The CTH model was shown to match the deformation measured in the tests, as well as the temperature profile observed in the microstructure. That is, CTH was capable of matching all of the major aspects of gouging that were experimentally noted in the hypervelocity gouging tests.

Additionally, the results from the hypervelocity gouging tests were shown to match between experiment, the one-dimensional penetration theory, and CTH simulations for the cases examined. This established confidence in both the one-dimensional theory and the CTH results.

In this manner, the use of CTH to model hypervelocity impacts with the new material constitutive model was validated. The CTH simulations were able to capture the deformation of the material and the associated temperature profiles resulting from the plasticity.

IX. Simulation of the HHSTT Gouging Scenario

Having developed a specific material constitutive model for the materials in use at the HHSTT, and establishing the accuracy of CTH simulations of hypervelocity impacts, the model of the HHSTT gouging scenario was evaluated. Similar to results from previous researchers, three major techniques are used to generate gouging impacts in CTH.

The first case examined was having the sled shoe come down with a given vertical velocity while traveling down the rail at 3 km/s. Impacts velocities on the order reported as the conditions during sled runs did not generate gouging, but did show some small amount of plasticity in the coating and did generate a small zone of localized heating beyond the austenizing temperature. To create gouging of the magnitude seen in the metallurgically examined gouge from the HHSTT, a much higher vertical impact velocity was required. If that velocity were applied, gouging and the associated temperature profile that matches the HHSTT gouge were created. Additionally, reducing the contact surface area of the shoe led to increased gouging. Finally, a heated shoe scenario was presented and shown to not significantly modify the overall results (out to 20 μs). Throughout this analysis, the one-dimensional penetration theory matched the resulting penetration depths fairly well.

The second case studied was to create an angle of incidence between the shoe and rail at impact. This causes the leading edge of the shoe to contact the rail prior to the remainder of the shoe. Using the maximum angle of incidence allowed by the nominal shoe gap, the geometry of the test sled, and the HHSTT rail seam tolerances, the impact generated a wearing (nongouging) impact. This wearing impact, however, generated enough local plasticity and temperature into the depth of the rail as to match the metallurgical results seen from the in-service (but visually undamaged) rail from the HHSTT. Increasing the impact angle (to one that cannot occur unless the rear shoes fail to hold the sled down on a flat rail or that can occur within the rail seam tolerances) resulted in gouging that matched the gouge examined from the HHSTT. Both the amount of damage and the generated temperature profiles match the observed gouge characteristics.

The final case explored was the collision of the shoe into a rail discontinuity, typically the result of rail misalignment at the section end. Several rail discontinuities heights were examined. For a rail discontinuity of the nominal thickness of the coatings used at the HHSTT, the resulting impact generated a wear-type response. This impact created a temperature profile that would modify the microstructure of the rail to a depth of the magnitude seen in the in-service rails examination. Increasing the discontinuity height to 0.02 cm resulted in the type of gouging that could create the HHSTT gouge specimen. This height is only 32% of the allowable rail misalignment tolerance. This particular simulation was unique in that it also predicted the creation of melted material, which has been observed experimentally. A slightly larger misalignment height was also simulated, with similar results. A full study of varying rail discontinuity heights and face angles was conducted to determine the relationship between these parameters and the occurrence of gouging. A power-law fit was made to the threshold values, which created wearing damage and prevented gouging. Current HHSTT tolerances were shown to be within the range of values that can lead to hypervelocity gouging. Therefore, rail discontinuities that could exist in the field have been shown to create hypervelocity gouging in the CTH simulations. A new tolerance criterion was developed to ensure that gouging would not occur in future test runs.

In making a judgment as to the probable cause of the hypervelocity gouging, the gouges predicted by CTH from the various techniques were compared qualitatively to the geometry seen at the HHSTT. The rail-discontinuity-type initiation of gouging was shown to be the most likely cause of hypervelocity gouging. In addition, the CTH model predicts gouging begins at values of rail misalignment that are within the current accepted tolerance standards at the HHSTT.

X. Conclusion

In this study, several advances were made in the field of understanding and modeling hypervelocity gouging.

An experimental examination of the gouges from the HHSTT led to the discovery of a thermal history within the gouging process. This thermal evidence was also found in rails that had not experienced visual damage. The ability of the rail steel to conduct the plastically generated heat rapidly enough to create the observed microstructures was verified.

A series of experimental tests were conducted to formulate the constitutive models for VascoMax 300 and 1080 steel that had not been previously accomplished. These models were validated using a series of experiments at various strain rates and conditions. Additionally, the relative effect of the HHSTT coatings on friction was shown.

A scaled hypervelocity gouging test that successfully recreated this type of gouging in controlled conditions was created. Associated with this effort, a one-dimensional penetration theory that can estimate penetration depth with a minimum of material information and no additional experimentation was developed.

The CTH modeling of hypervelocity impacts was validated. Specific gouge characteristics, such as deformation and temperature profiles, were shown to match experimental measurements. These simulations relied upon the developed material constitutive models and the embedded nonequilibrium thermodynamic-based (experimental) equation-of-state information.

Finally, with the validated material flow models developed for the HHSTT scenario, a CTH model was utilized to determine that a rail discontinuity is the probable cause of hypervelocity gouging and that the CTH simulation recreates major aspects of the gouging experimentally observed. A new rail-alignment criterion, based on numerous simulations, is proposed that should mitigate future incidences of hypervelocity gouging.

References

[1] Cinnamon, J. D., and Palazotto, A. N., "Further Refinement and Validation of Material Models for Hypervelocity Gouging Impacts," *AIAA Journal*, Vol. 46, No. 2, 2008, pp. 317–327.

[2] Cinnamon, J. D., and Palazotto, A. N., "Analysis and Simulation of Hypervelocity Gouging Impacts for a High Speed Sled Test," *International Journal of Impact Engineering*, Vol. 36, Issue 2, 2009, pp. 254–262.

Sample Scaled Hypervelocity Impact CTH Input File

```
*eor* genin

Oblique Impact: V300/1080, v=2225, angle=10, ux=2192.2, uy=-386.37 m/s

* All output from CTHgen will have this title.
*
* This is an impact scenario of a Vascomax 300 projectile
* hitting a 1080 steel target.
*
control
  mmp3
  ep
  vpsave
endcontrol
*
*________________________________________________________________________
*
mesh
  block 1 geom=2dr   type=e        * This section defines the mesh, not
                                     the cylinder or block.
                                   * 2dr = 2D rectangle (plain strain) -
                                     Use 2dc for axisymmetric.
                                   * e = Eulerian solution, the only
                                     current option

     x0=0.0                        * x0 = left starting value of x.
       x1 n=10 w=1.0 dxf=0.100     * x1,x2 = x subzone for defining
                                     different meshes along x.
       x2 n=50  w=0.5 dxf=0.010    * ALL DIMENSIONS MUST BE IN cm!
       x3 n=220 w=1.1 dxf=0.005    * n = # of cells; w = section width;
       x4 n=40  w=0.4 dxf=0.010    * dxf = dx first; dxl = dx last
                                     (optional)
       x5 n=20  w=2.0 dxf=0.100
     endx
     y0=-4.0                       * y has same inputs as x
       y1 n=20  w=2.0 dyf=0.10     * y0 = bottom starting position.
       y2 n=140 w=1.4 dyf=0.01
       y3 n=240 w=1.2 dyf=0.005
       y4 n=40  w=0.4 dyf=0.10
     endy
  endb
endmesh
*
*________________________________________________________________________
*
```

```
insertion of material
  block 1
     package rod          * Insert rectangular rod and rotate
     material 1
     numsub 100

     xvel 2192.2e2        * Change only the first number, leave 'e2'
                            this converts m/s to cm/s
     yvel -386.37e2

     insert box           * This is where you input the cylinder.
       pl 0.0 0.0         * Only model 1/2 of cylinder, and then
                            "mirror" in plotting.
       p2 2.225 0.55      * Format: pl = bottom center point of
                            cylinder
                          * p2 = top right hand corner of cylinder.
                          * Change pl and p2 to define the size of rod.
     endinsert

     delete circle
       center 2.225 0.275
       radius 0.275
     enddelete
  endpackage

  package tip             * Insert circular rod tip
     material 1
     numsub 100
     xvel 2192.2e2
     yvel -386.37e2

     insert circle
       center 2.225 0.275
       radius 0.275
     endinsert
  endpackage

  package target          * Insert Rail Material
     material 2
     numsub 100
     insert box
       pl 0.0 -4.0
     p2 5.0 -0.02
     endinsert
  endpackage

  package coating
     material 3
     numsub 100
     insert box
       pl 0.0 -0.02
       p2 5.0 0.0
     endinsert
  endpackage

 endblock
endinsertion
*
*__________________________________________________________________________
*
edit
  block1
  expanded
  endblock
endedit
*
*__________________________________________________________________________
```

```
*
eos                            * Eq Of State
* Information for metals
   MAT1 SES STEEL_V300         * Sesame Eq of State tables, limited
                                 material selection.
   MAT2 SES IRON               * Iron = closest material available to 1080
                                 steel.
   MAT3 SES IRON               * iron oxide=MAT3 SES IRON, epoxy=MAT3 MGR
EPOXY_RESIN1
endeos

epdata                         * Elastic/Plastic data
  mix 3                        * mix 3 = normalized vol avg yield strength

   matep 1 ZE USER             * ZE coefficients for VascoMax 300.
      aze=1.42e10              * Dynes/cc
      nze=0.289
      c1ze= 4.0e10             * Dynes/cc
      c2ze=0.0
      c3ze=79.0                * 1/eV
      c4ze=3.0                 * 1/eV
      c5ze=0.266e10            * dynes/cc
      poisson= 0.283

   matep 2 ZE USER             * ZE coefficients for 1080 steel.
      aze=0.825e10             * Dynes/cm^2
      nze=0.289
      c1ze=4.0e10              * Dynes/cm^2
      c2ze=0.0
      c3ze=160.0               * 1/eV
      c4ze=12.0                * 1/eV
      c5ze=0.266e10            * dynes/cc
      poisson= 0.27

   matep 3
      Yield = 0.1e8            * iron oxide=0.1e8, epoxy=0.5e7
      poisson=0.46             * iron oxide=0.46, epoxy=0.26
  vpsave
  lstrain
endep
*
*______________________________________________________________________
*
*tracer
                    * Tracer input, starting at x1,y1 to ending x2,y2.
                    * n=number of tracers to distribute including
endpoints.
         * Projectile boundary
*  add 1.0 .01 to 3.0 .01 n=50
*endt
*
*______________________________________________________________________
*
*eor* cthin

Oblique Impact: V300/1080, v=2225, angle=10, ux=2192.2, uy=-386.37 m/s

control              * Defines sys parameters for execution.
  mmp3               * mmp = multiple pressures and temps in mixed
material cells.
  tstop = 15.0e-6    * tstop = Problem stop time in seconds.
  cpshift=600.       * cpshift allows extra time for CTH to right data.
endc
*
*______________________________________________________________________
*
```

```
Convct                          * Convection control input.
 convection=1                   * Convect internal energy based on
                                  energy density and mass
                                * density. Discards kinetic energy.
 interface=high_resolution      * interface tracker, high_res
                                  recommended for 2D.
endc
*
*__________________________________________________________________
*
fracts               * Fracture Strength
 pfrac1=-2.2e10      * pfrac# = fracture pressure/stress of material #.
 pfrac2=-1.4e10
 pfrac3=-0.1e8       * iron oxide=-0.1e8, epoxy=-0.5e7
 pfmix=-2.0e10       * pfmix = fracture stress or pressure in cell with
                       mixed mat, no voids
 pfvoid=-2.0e10      * pfvoid = fracture press or stress in cell with
                       void.
endf
*
*__________________________________________________________________
*
edit
  shortt
    tim 0.0, dt=1.0         * Restart data will be written every 'dt'
seconds.
ends

longt
  tim 0.0, dt=1.0           * Restart data will be written every 'dt'
seconds.
  endl

  plott .
    tim 0.0 dt=0.5e-6       * cthplot data is written to restart file
every
                            * 'dt' seconds starting at 'tim'.
  endp                      * Beware - Restart file size limited to 2GB.

  histt
    tim 0.0 dt=0.5e-6       * History data will be written to hcta every
    htracer all             * 'dt' seconds. Data will be written for all
                            * tracer points.
  endh

ende
*
*__________________________________________________________________
*
boundary
     bhy                          * rigid boundaries all around
       bl 1
         bxb = 1, bxt = 1
         byb = 1, byt = 1
       endb
     endh
endb
*
*__________________________________________________________________
*
vadd                              * Apply vel to keep gouge in mesh
  block=1
  tadd=0.0
  xvel=-2192.2e2
endvadd
```

```
*
*_____________________________________________________________________
*
cellthermo
 mmp3                      * This was recommended by Eglin and appears
                           * to give good results as well.
 ntbad 500000000
endc
*
*_____________________________________________________________________
```

Appendix B

Sample Sled Simulation CTH Input File

```
*
*eor* cgenin
*
* cthgen input for Gouge simulation
*
*
*           _______________
*          |    ------>    |
*          |    |          |
*          |    v          /
* -------------------------------
*
*
* vx=varies, vy=-1 m/s V300 Steel Slider, 1080 Steel Rail, No Atm.
* No Slide line. mix=1 frac=1 Rounded corner.
* Added mass on top to simulate sled mass
*
 -1 m/s, 3.0 km/sec, 6 mil coating, .0020 cm mesh

control
  mmp3
  ep
  vpsave
endcontrol

*_/_/_/_/_/_/_/_/_/_/_/_/_/_/_/_/_/_/_/_/_/_/_/
* MESH AND BLOCK DEFINITION SET
***********************************************
* geom=2DR(rectangular x,y)
* geom=2DC(cylindrical x=radius, y=axis)
* geom=3DR(rectangular x,y,z)
* type=e (Eulerian)
* x#=coordinate range for plot
* y#=coordinate range for plot
* dxf=width of first cell in the region
* dxl=width of last cell in the region
* n=number of cells added in this region
* w=total width of this region in centimeters
* r=ratio of adjacent cell widths
*_/_/_/_/_/_/_/_/_/_/_/_/_/_/_/_/_/_/_/_/_/_/_/
mesh
 block 1 geom=2dr type=e
   x0=0.0000
     x1 w=15.00 dxf=5.00 dxl=0.002
     x2 w=3.000 dxf=0.0500 dxl=0.002
     x3 w=12.000 dxf=0.050 dxl=0.002
   endx
   y0=-4.000
     y1 w=2.200 dyf=0.100 dyl=0.0500
```

```
     y3 w=1.000 dyf=0.050 dyl=0.002
     y4 w=1.600 dyf=0.002 dyl=0.002
     y5 w=2.200 dyf=0.002 dyl=0.050
     y6 w=5.000 dyf=0.0500 dyl=0.100
     y7 w=35.500 dyf=0.100 dyl=10.00
    endy
  endblcck
endmesh

insertion of material
  block 1

package rail
      material 2
      numsub 100
      xvel 0.0
      yvel 0.0
      insert box
      pl 0.0 -0.01
      p2 50.0 -4.0
      endinsert
endpackage

package coating
      material 1
      numsub 100
      xvel 0.0
      yvel 0.0
     insert box
     pl 0.0 0.0
     p2 21.0 -.01
     endinsert
endpackage

package coat2
      material 1
      numsub 100
      xvel 0.0
      yvel 0.0
      insert triangle
      point 21.0 -0.01
      point 23.0 0.0635
      point 30.0 -0.01
      endinsert

      delete triangle
      point 21.1 -0.01
      point 23.0 0.04826
      point 28.5 -0.01
      enddelete
endpackage

package rail2
      material 2
      numsub 100
      xvel 0.0
      yvel 0.0
      insert triangle
      point 21.1 -0.01
      point 23.0 0.04826
      point 28.5 -0.01
      endinsert
endpackage

package shoe
      material 3
```

```
      numsub 100
      temperature = 2.53575e-2 * eV = 70F
       xvel 3.0e+5
       yvel -1.0e+2
     insert box
        p1 0.300, 0.0
        p2 20.62, 2.54

      endinsert
      delete circle
        center 20.42, 0.2
        radius 0.2
      enddelete
      delete box
        p1 20.42, 0.2
        p2 20.62, 0.0
      enddelete
    endpackage

      package sledsim
      material 4
      numsub 100
      temperature = 2.53575e-2 * eV = 70F
       xvel 3.0e+5
        yvel -1.0e+2
      insert box
         p1 0.300, 43.14
         p2 20.62, 2.54

       endinsert
     endpackage

     package sliderround
       material 3
       numsub 100
        temperature = 2.53575e-2 * eV = 70F
        xvel 3.0e+5
        yvel -1.0e+2
        insert circle
            center 20.42, 0.2
            radius 0.2
      endinsert
    endpackage
  endblock
endinsertion

edit
  block 1
  expanded
  endblock
endedit

*_/_/_/_/_/_/_/_/_/_/_/_/
* TRACER DEFINITION SET
*_/_/_/_/_/_/_/_/_/_/_/_/
*tracer
  *Target guider
*  add 20.40, -0.02 to 26.70, -0.02 n=10
  *Target slider
*  add 00.40, 0.02 to 20.40, 0.02 n=10
  *Target coating
*  add 20.40, -0.00725 to 26.70, -0.00725 n=10
  *Target slider boundary layer
*  add 00.40, 0.003 to 20.40, 0.003 n=10
endt
```

```
*_/_/_/_/_/_/_/_/_/_/_/_/_/_/_/_/_/_/
* EQUATION OF STATE DEFINITION SET
*_/_/_/_/_/_/_/_/_/_/_/_/_/_/_/_/_/_/
eos
*   MAT1 SESAME=GE1_RP EOS=7662* Reactive Graphite Epoxy
*   MAT1 MGR EPOXY_RESIN1
*   MAT1 SESAME=EPOXY EOS=7602* FEOS='seslan'
*   MATI SES EPOXY
    MAT1 SES GREPXYI for epoxy
*   MATI SES IRON for iron oxide
    MAT2 SES IRON
    MAT3 SES STEEL_V300
    MAT4 MGR PLATINUM
endeos

epdata
   vpsave
   lstrain
   mix=3
   matep=1  *Epoxy Glider Coating
     poisson 0.46
     yield 1.0e8

   matep=2 ZE USER * 1080 Steel rail
     aze=0.825e10   * dynes/cm^2
     nze=0.289
     c1ze=4.0e10    * dynes/cm^2
     c2ze=0.0
     c3ze=160.0     * 1/eV
     c4ze=12.0      * 1/eV
     c5ze=0.266e10  * dynes/cm^2
     poisson 0.27

matep=3 ZE USER       * VascoMax 300 shoe
   aze=1.42e10         * dynes/cm^2
   nze=0.289
   c1ze=4.0e10         * dynes/cm^2
   c2ze=0.0
   c3ze=79.0           * 1/eV
   c4ze=3.0            * 1/ev
   c5ze=0.266e10       * dynes/cm^2
   poisson 0.283

matep=4
   poisson .2
   yield 10e10
ende

**************************************************************************
*eor* cthin
-1 m/s, 3.0 km/sec, 6 mil coating, .0020 cm mesh

*_/_/_/_/_/_/_/_/_/_/_/_/_/_/_/_/_/_/_/_/_/_/_/_/_/_/_/_/_/
*CONTROL DATA SET
*******************************************************
* mmp enables multiple material temps
* and pressures in mixed cells
* frac=1 changes fracture default
********************************
* CELL THERMODYNAMICS INPUT SET
********************************
* cellthermo
*   dtmax = max temp difference allowed in mixed cells
*   mmp distributes volume and energy based on volume
*   fractions of material in the cell
*   mmp=default
```

```
*  mmp1=same as mmp except uses new logic
*      and distributes volume and energy
*      based on volume fraction cubed divided
*      by mass of material in the cell
*  mmp2=allocation of work done on the cell is
*      dependent on material compressibility
*      and allows pressure relaxation between materials
*      in a cell
*  mmp3=new default that produces best results, see manual
*endc
*_/_/_/_/_/_/_/_/_/_/_/_/_/_/_/_/_/_/_/_/_/_/_/_/_/_/_/_/
control
   mmp3
   tstop = 40.00E-6
   nscycle=55000
   rdumpf=991800.
   cpshift=600.
   ntbad=99999999
*<<<<<<<<<<<<<<<<<<<<<<<<<<<<<<<<<<<<<<
* Courant condition multiplier
*<<<<<<<<<<<<<<<<<<<<<<<<<<<<<<<<<<<<<<
   dtcourant=0.6
endc

Convct
   convection=1
   interface=high_resolution
endc

fracts
pfrac1=-1.0e8
pfrac2=-2.0e10
pfrac3=-2.5e10
pfrac4=-1.2e10
pfmix=-1.20e10
pfvoid=-1.20e10
endfracts

edit
   shortt
     time = 0.0 , dt=1.0e-6
   ends
   lonqt
     time = 0.0e0 , dt = 1.0
   endl
   plott
     time 0.0e-6 dtfrequency 0.5e-6
   endp
   histt
     time 0.0e-6 dtfrequency 0.5e-6
     htracer all
   endhistt
ende

*_/_/_/_/_/_/_/_/_/_/_/_/_/_/_/_/_/_/_/_/_/_/_/_/_/_/_/_/
* Hydrodynamic Boundary Conditions
*********************************************************
* 0=symmetry
* 1=sound speed based absorbing
* 2=extrapolated pressure with no mass allowed to enter
* 3=extrapolated pressure but mass is allowed to enter
*_/_/_/_/_/_/_/_/_/_/_/_/_/_/_/_/_/_/_/_/_/_/_/_/_/_/_/_/
boundary
  bhydro
    block 1
       bxbot = 1 , bxtop = 1
```

```
        bybot = 1 , bytop = 1
     endb
   endh
endb

*_/_/_/_/_/_/_/_/_/_/_/_/_/
* Heat conduction inputs
*_/_/_/_/_/_/_/_/_/_/_/_/_/
*heatconduction
* MAT1 TABLE=3
* MAT2 TABLE=1
* MAT3 TABLE=2
*endh

* DEFTABLE=1 * 1080 STEEL
* T(eV) k(erg/s/eV/cm)
* 1.4684e-3 4.7700e10
* 1.0377e-2 4.8100e10
* 1.9090e-2 4.5200e10
* 2.7900e-2 4.1300e10
* 3.6711e-2 3.8100e10
* 4.5521e-2 3.5100e10
* 5.4332e-2 3.2700e10
* 6.3142e-2 3.0100e10
* 7.1953e-2 2.4400e10
* 8.9574e-2 2.6800e10
* 1.1111e-1 3.0100e10
* endd

* DEFTABLE=2 * VascoMax 300 Steel
 *T(eV) k(erg/s/eV/cm)
* 3.6711e-3 2.4715e10
* 1.4684e-2 2.7424e10
* 2.9369e-2 2.9794e10
* 3.9158e-2 3.0132e10
* endd

* DEFTABLE=3 * Epoxy
 *T(eV) k(erg/s/eV/cm)
* 3.6711e-3 6.5e8
* 1.4684e-2 6.5e8
* 2.9369e-2 6.5e8
* 3.9158e-2 6.5e8
* endd

*_/_/_/_/_/_/_/_/_/_/_/_/_/_/_/_/_/_/_/_/_/_/_/_/_/_/_/_/_/
* Added velocity to maintain gouging in view
*_/_/_/_/_/_/_/_/_/_/_/_/_/_/_/_/_/_/_/_/_/_/_/_/_/_/_/_/_/

*vadd
* block=1
* tadd=0.0
* xvel=-1.08333e+5
*endvadd

*mindt
* time=0. dt=1.e-12
*endn
maxdt
   time=0. dt=.01
endx

cellthermo
mmp3
endc
```

Index

Supporting Materials

Many of the topics introduced in this book are discussed in more detail in other AIAA publications. For a complete listing of titles in Progress in Astronautics and Aeronautics, as well as other AIAA publications, please visit http://www.aiaa.org.